T0342074

The Scientific Bases
of Human Anatomy

Advances in Human Biology

Series Editors:
Matt Cartmill
Kaye Brown
Boston University

Titles in this Series

The Scientific Bases of Human Anatomy

CHARLES OXNARD
Emeritus Professor, University of Western Australia

WILEY Blackwell

Published by John Wiley & Sons, Inc., Hoboken, New Jersey
Published simultaneously in Canada

Library of Congress Cataloging-in-Publication Data:

Oxnard, Charles E., 1933- author.
  The scientific basis of human anatomy / by Charles Oxnard.
     pages cm
  Includes bibliographical references and index.
  ISBN 978-0-471-23599-6 (cloth)
  1. Human anatomy. I. Title.
  QM23.2.Q965 2015
  612–dc23
                                                    2015006925

Cover image: © Charles Oxnard

Set in 10/12.5pt ITCGaramondStd by SPi Global, Chennai, India.

Printed in Singapore by C.O.S. Printers Pte Ltd

1   2015

For Eleanor
I was a studious student: I married the Medical Librarian
She has been in my work and my life for nearly 60 years
Willingly helping me cross three continents

Contents

Foreword

To readers who think of human anatomy as a petrified science, in which all the facts are established and all the big questions have been answered, Charles Oxnard's new book will come as a surprise. *The Scientific Bases of Human Anatomy* is unlike any other book on the subject. In reading it, you will come to perceive your own body and the bodies of others in a dramatic new light, as the culmination of a story: the narrative of the journey that our bodies have taken to become human.

In other books of human anatomy, the bold outlines of this story are washed away in an inundation of facts. Anatomical pedagogy has traditionally relied on mnemonics and rote memorization to support learning and recall of this flood of detail. Oxnard's approach uncovers the deficiencies of this tradition and overcomes them. His book shows us a way around reliance on memorization and mnemonics by concentrating on how human bodies come into existence. The analytical system that he uses relies on identifying recurring patterns and tracing these back to the processes that formed them. In Oxnard's words,

> The method that I have developed, over the years is, in contrast to 'the naming of the parts', an holistic integrative approach to human structure. It involves understanding that the structural details of the human body are the end results of a series of biological processes. These produce anatomical pattern due to:
>
> - change from differentiation and growth **over developmental time;**
> - diversification through comparisons of different forms **living at the same time;**
> - adaptation to mechanical and other factors **during functional time;**
> - interaction between body and brain, brain and body, **through mind time**, and
> - innovation in structure resulting from evolution **during deep time.**

Oxnard escorts his readers on a guided exploration of what he identifies as "the principles of body construction from their beginnings" – a tour of the human body factory, in the company of a most engaging guide who reveals and explains how bodies are made and how they work. This exploration is grounded in current ideas drawn from a wide arc of biological sciences, ranging from genomics and neuroscience to the latest findings in comparative anatomy. In the hands of a less skilled and knowledgeable writer, the confluence of all these "developmental, comparative, adaptive, integrative (and) evolutionary" ideas would have left the reader lost in a wilderness of unrelated notions. But Oxnard's mastery of biology and passion for anatomy hold the reader's journey on a steady course, relieved by a few delightful side excursions and enlivened by his unique and accessible narrative style. Profound, strong, sure, sometimes poetic and even beautiful, Oxnard's distinctive voice will remain with his readers long after they have finished this book.

Oxnard is careful to point out that *The Scientific Bases of Human Anatomy* is not a textbook of anatomy. As he emphasizes in Chapter 1, there is nothing here to memorize, and it is not his intention to prepare the reader for a test based on the "naming of the parts." Reading Oxnard will not obviate the need for professionals to acquire this sort of detailed anatomical knowledge, but it will both lighten and illuminate that task. His book conveys lasting images of how bodies come into being and function, which will help students organize those details in ways that make fundamental sense. For teachers of anatomy, the new insights and ways of thinking laid out in the following pages may serve to rekindle the spark of inquiry that drew them to the topic in the first place. For uninitiated readers with no professional interest in anatomy, the book will raise the curtain on a theater of the mind in which they will come to care about the making of bodies. In this book, Oxnard seeks to lay down new modes of understanding and thinking about the formation of the human body, as both process and product, for students, teachers, researchers, and others. All of his readers will henceforward see the human body in a novel and deeply enlightening way. We are honored and delighted to include his book in the *Advances in Human Biology* series, and to welcome its readers to the New Anatomy.

KAYE BROWN AND MATT CARTMILL

Preface

How it started: I was trained as an old fashioned medical anatomist, a physician who chose to specialize in anatomy; today a species possibly extinct, probably obsolete. Thus, though I just escaped being a student in a medical anatomy course of one thousand hours in two years, I nevertheless took a medical anatomy course of six hundred hours in five terms. I dissected the entire human body and brain and I taught in this mode for six years. I was later involved in teaching a medical anatomy course in two quarters (still with dissection of the entire body), and then a course in one quarter (with dissection of the entire body!). Finally, I was involved in teaching a course that was not entitled anatomy at all, but that contained just enough anatomy to understand physiological systems such as the cardiovascular system, and simple clinical problems. It contained no regional anatomy (though perhaps one third of medical conditions involve complex anatomical regions like the back of the abdomen) and no human dissection (though a small aliquot of students were permitted to take an elective dissection course in which they dissected a single region of the body of their own choosing). These gradual reductions of anatomical teaching meant that my own teaching became better and better – the smaller the amount of teaching, the higher the quality of teaching!

The researches that I carried out concomitantly with this medical student teaching led me to undertake undergraduate and graduate courses in the scientific bases of human anatomy. For, if one wants to attract graduate and post doctoral students, it is wise, in the American system, to provide graduate level courses in one's own direct discipline. But how does one transmute that kind of medical human anatomy into science-based human anatomy?

The transmutation: This was achieved, without any difficulty, by serendipity. Thus, at a very early stage I had a colleague, a technician in the British university system, who had become so fascinated by anatomical research that he undertook some of it himself. The result: he published a series of papers; these papers could be used to apply for what was called an 'official degree'. As he had no previous degree, it had to be a Bachelor of Anatomical Sciences. Just as he thought it was all settled, he was then informed by the University Senate that he had also to take a written examination in Anatomical Sciences. He was devastated. Extremely bright though he was, he felt that he had had no formal education, and that there was no way he could write the essays for such a requirement.

We conferred. I agreed to give him one-on-one tuition on the scientific bases of human anatomy (which in my book included vertebrates, chordates, and even more) for an entire year. I am most pleased to say that he passed, and went from Junior Technician to, eventually, Senior Lecturer, a transition that was generally almost impossible in the British university system at that time. He was delighted; we remained very close scientific colleagues until the end of his life.

Though this episode did so much for him, it did, perhaps, very much more for me. It meant that I had to look at medical anatomy and reinterpret it through its scientific bases. Of course, at that time, the late fifties, early sixties, of the 20th century, that task could not be fully achieved. The modern genetic, developmental, comparative, functional and evolutionary bases had just not been taken far enough. However, ever since that time, a series of other institutions, students and colleagues, have provided me with the materials, ideas and stimulation, to keep working on that concept.

Chicago, the next step: For me, following Anatomy at the University of Birmingham in the fifties and early sixties, came Anatomy at the University of Chicago in the late sixties and seventies. Although originally a department that taught human anatomy to medical students, it had become transformed under Ronald Singer into a department that included a group of individuals interested in organismal, developmental, functional and evolutionary biologies. I was one of Singer's appointments in that transformation.

My view of human anatomy was further colored by the other evolutionary biologists who were already there: Singer himself, David Wake, James Hopson, Eric

Lombard, Len Radinsky, Lorna Straus, Russ Tuttle, and Leigh Van Valen, together with all the other anatomists in the cell and molecular biology areas. These ideas were further strengthened by discussions with the research student cohort of those days, in alphabetical order: Gene Albrecht, Matt Cartmill, Rebecca German, Walter Greaves, Paul Heltne, Doug Lay, Betty Jean Manaster, Jane Peterson, Jim Shafland, Jack Stern, and Richard Wassersug. These individuals, all interested in human anatomy, all preparing themselves for teaching anatomy to medical students, were, at the same time, interested in the scientific underpinnings of anatomy in organismal, developmental, functional and evolutionary modes. These ideas were further influenced by distinguished colleagues in department outside Anatomy: Stuart and Jean Altmann, Jack Cowan, Al Dahlberg, Richard Klein, Dick Lewontin, Lyn Throckmorton among others. Finally, these ideas were influenced by some of the visitors that we had in that Anatomy Department: Fred Bookstein, George Lauder, La Barbera, Rene Thom, later Neil Shubin (unfortunately after my time, but an influence on me through his book "Our Inner Fish"), and many others who visited Chicago in those days.

Translation to Los Angeles: My translation to the University of Southern California, though initially primarily as graduate dean and Professor of Anatomy, also included appointment as University Professor. These positions allowed continued research and teaching in Biology and Anatomy using the above principles. They involved further work with Gene Albrecht (who had moved there while I was Dean, but also involved a series of other colleagues: Fred Anapol, Bruce Gelvin, Joe Miller, Pete Lestrel, Brad Blood, Sherry Gust, Artyan Hsu, and others. Once again, I continued applying the underlying biological science to my teaching of medical anatomy (though the course was organized and run by Gene Albrecht); all these colleagues contributed.

This was also an especially seminal time for the medical students. They had started to realize that it might be important that they have a research side to their lives. Partly I think this was because of the general interest of extremely bright students; partly, however, it was because they could see that the official, administrative and accounting pressures of the 'new medicine' might need to be leavened by other interests such as research and teaching. Thus, I was approached by a group of some 25 medical students lead by Dan Zinder; they already held undergraduate engineering degrees. They knew my researches included engineering methods. They wanted me to be the director of an 'Engineers in Medicine Group'. Then I was approached by another group of medical students (leaders July Graylow and Hugh Allen) who just wanted to do research in one or other of the laboratories around the medical school; so I became director of a 'Medical Student Research Group'. These students especially participated in the Western Medical Students Research Forum. Finally, perhaps partly as a result of these endeavors, I became the first director (though only for one year as a result of my next move) of the new 'MD/PhD program' in the USC medical school.

Termination (!) in Western Australia: My final move to the University of Western Australia (UWA) has allowed me to continue my way of looking at anatomical teaching. This was all the more powerful at UWA because, in addition to medical student teaching, we had a full three year program for undergraduates in Human Biology, a strong 4th year Honors research undergraduate program, several Masters programs, as well as standard doctoral studies in aspects of Human Biology (and later also in Forensic Science). But most of all, however, in Anatomy and Human Biology the undergraduate courses already embodied some of these ideas. They were promoted first by David Allbrook and then masterminded by Leonard Freedman. They were further carried forward by many of the academic staff, but especially by Neville Bruce. As a result, I found myself entering a school where the ideas that I had were already in full swing, some, perhaps, well beyond my own thoughts. It was a marriage made in heaven.

These ideas were again yet further stimulated by a series of graduate students, colleagues, and others (but now academics of various ranks) including, but not limited to, Vanessa Hayes, Alanah Buck, Nick Milne, Elizabeth Pollard, Ken Wessen, Robert Kidd, Willem de Winter, Pan Ruliang, Jens Hirschberg, Algis Kuliukas, Dan Franklin, and, most recently, Sara Flood, Warren Mitchell and Sally Stevens (not a student of ours, but a most capable teacher). They all taught in our courses, carried out research in the overall area, and continued to help me after they became academics. One recent student, Sarah Flood (now an Assistant Professor) made especially strong but appropriate criticisms of drafts of this book.

In Retirement: Since then, during 'so-called' retirement, I have been enabled to continue with both my researches and these ideas on teaching the scientific bases of human anatomy. In part this is because the University of Western Australia, through the School of Anatomy, Physiology and Human Biology, and the Forensic Science Centre, continues to permit me to have my research office, space for my graduate students and post-docs, and internal research funds. In part, also, it is because the Australian Research Council and the Medical and Health Infrastructure Fund of Western Australia have continued to provide me with project and infrastructure research funding during the whole of my retirement (my latest ARC and MHRIF grants will not expire until 2015).

In large part, further, it is because, during my retirement, a series of overseas appointments, grants and colleagues have also supported my ideas. These have included, in the last few years, a three year part-time Leverhulme Professorship at University College, London, two Leverhulme Research Grants, a BBSRC Research Grant, and Marie Curie Research and Research Training Funds (all in the UK). These have been achieved through the help of UK colleagues: Paul O'Higgins (first at University College, London, and later at the University of York), Robin Crompton, (University of Liverpool), and Michael Fagan, (University of Hull). Important insights have also been provided by interactions with other faculty, graduate students, and post-docs in York, Hull, Liverpool, Dundee, Zurich and Vienna.

I must also record my indebtedness to a series of technicians and artists who have supported my work on all three continents: they include Bill Pardoe (Birmingham), Joan Hives (Chicago), Erika Oller (Los Angeles), and Martin Thompson, Rebecca

Davies and Sue Hayes (Perth). Their hands are clearly evident in the improvements they made to my initial teaching illustrations.

I am grateful to Matt Cartmill and Kaye Brown, the academic editors of the Foundations of Human Biology book series at Wiley Academic Press in which this volume plays a part. I also owe thanks to the various editorial staff at Wiley. All have been particularly forgiving of the long time this book has been in progress. Matt himself, once a person I taught, has taught me much through his reviews in these final stages.

Finally, four individuals have been with me most of the way.

The first was Tom Spence who died in 1997. He started as a junior technician in Anatomy in Birmingham at the end of WWII, but made the transition, extremely difficult in England at that time, to academic ranks. He eventually retired in 1980 as Senior Lecturer. I had the good fortune to be able to visit him most years in his retirement even though I was then located in the USA. I remember so much our hours of discussion, and the teas (and for me a beer) supplied by his wife Joan, to keep us going. It was, however, his need, while still a technician, for an understanding of the scientific bases of human anatomy, for help to enable him to pass an examination for his Official BSc Degree, that first stimulated many of the ideas underlying this book.

Second, ever since my own entry into Anatomy in 1952, I have been in frequent contact with Peter Lisowski, first at the University of Birmingham, then at Haile Selassie University, Ethiopia, later when he was Head of Anatomy at the University of Hong Kong, and finally at the University of Tasmania, Hobart. Peter and I taught together, examined together and published together. In particular Peter Lisowski was the academic anvil upon which my ideas on the scientific bases of human anatomy were hammered out. Though we were both old-fashioned medical anatomists by training, we both had ideas that were over and beyond the traditional teaching of medical anatomy. He died on 11 January 2007; Eleanor and I were grateful to be with him and Ei Yoke, his wife, only a few days before he slipped away.

The third individual was Len Freedman who spent many years in building up the idea of Human Biology as the science underlying the human condition at the University of Western Australia. I had first met him during Chicago days when he was at the University of Wisconsin. It was in Wisconsin that he was first enabled to elaborate his ideas on Human Biology, but it was on his later appointment at the University of Western Australia (UWA) that, with the total support of his professor at that time, David Allbrook, and of Neville Bruce, and the other academic staff at UWA, that he elaborated this form of Human Biology. Accordingly then, when I arrived at UWA many years later, I found a Human Biology niche already created, almost, as it were, for me. I have been with Len Freedman ever since, including his years in active retirement. He died a true academic's death, reading a book, in 2014 aged 90.

The final individual is, of course, my wife, Eleanor. Since 1954, when I first knew her, she has supported me and my work. Originally a medical librarian (I was a studious student, spent much time in the medical library) she helped in the bibliographies of my papers and books. Much more so, however, she has helped me through being willing to share our various migrations across the world. Sometimes the moves were initially tough, especially for her; but she fell in with them, made her own way, and remained indispensible to our work and life together. Even now we continue to travel the world yearly for family, collegial and research purposes.

This book has been such a wonderful thing to prepare. The new scientific bases of human anatomy are so much different from those with which I started 63 years ago. They are so unexpectedly exciting. Eleanor and I wish we were at the beginning of it all, and not near the end!

CHARLES OXNARD,
Claremont, WA, Australia
2015

CHAPTER ONE

A New System of Human Anatomy

1.1 Why a New System?

Another traditional human topographic anatomy book is definitely **not** needed. There are already a very large number of books on human topographic anatomy. They range from huge tomes attempting to lay out most of human anatomy to short summary books presenting the major facts in a pithy way. They form a spectrum from old books that present the anatomy as correctly as possible, including the many variations of normal, to new books that eliminate much information in order to present a simplified picture. They encompass the span from anatomy texts that are based upon anatomy from region to region (e.g. upper limb, thorax) of the

The Scientific Bases of Human Anatomy, First Edition. Charles Oxnard.
© 2015 John Wiley & Sons, Inc. Published 2015 by John Wiley & Sons, Inc.

body, to expositions that display anatomy in terms of systems (e.g. nervous, gastrointestinal, locomotor). There are picture volumes from anatomy coloring books of the principal features of the body, to major atlases showing, through hundreds of leader lines, every anatomical detail. More specifically, for the health professions, there are books emphasizing the anatomy relevant to exemplar clinical problems, and books of fully detailed clinical anatomy. There are even many ancillary texts showing specific parts of human anatomy (e.g. surface anatomy, imaging anatomy, anatomy for orthopedics, anatomy for nurses, anatomy for artists, and so on).

Almost without exception, however, these books present human topographic anatomy, in each of their different ways, as a road map of the human body to be memorized. As medical curricula have become increasingly crowded the modern anatomical road map has become more and more limited. Today the books (and most of the other sources) used by the medical student show only the motorways and freeways. The larger books have become little more than references for details of the 'lowways' and byways required as references for the medical specialist.

Why then is there a need for a new system of human anatomy?

Because scientific understanding makes human topographic anatomy, like any other science, derivative; it is something that can be handled, and not just an object for memorization.

Because introducing the underlying science can make the anatomy live in the imaginations of students.

Because scientific understanding of anatomy is an important background to many disciplines (such as physical anthropology, human biology, functional anatomy, primate, mammalian and even vertebrate anatomy and evolution) that need not just information but understanding of human anatomy.

Because an introduction to human anatomy via its scientific background is useful for a very large number of disciplines that are human-based but not especially anatomical (e.g. any of a large number of allied medical and health disciplines).

Because, most of all, major advances in several sciences underlying anatomy (such as genetics, developmental biology, molecular biology, growth, behavioral biology, neurobiology and evolutionary biology) now provide exciting new insights into the how and why of the structure of the human body.

As a result, especially of these last, understanding the scientific basis of human anatomy is of particular importance at this time. Of course, there has always been a scientific basis, even if it existed mainly in the minds of investigators, hopefully in the minds of teachers, even, if only rarely, in the minds of medical students. But the recent advances in certain other disciplines have new implications for understanding human anatomy.

1.1.1 What are These Advances?

A new developmental biology is resulting from deeper knowledge of genes and other developmental molecules, and especially through recent developments in understanding genetics and development from bioinformatics (e.g. in date order: Page and Holmes, 1998; Oxnard, 1983/1984; Hall, 1999; Larson, 2001; Moore, 2001; Twyman, 2001; Carlson, 2004; Shubin, 2008; Carey, 2012). These are all responsible for new understanding of developmental mechanisms and products. When I was a student there was a great gulf fixed between compound eyes in fruit flies and 'simple' eyes in humans. The old embryology had little further to say about this problem; it could show what existed but not how it came to exist. Who would have thought that closely similar genetic mechanisms and molecular processes would be discovered to exist for each; that a single new explanation in development would frame the multiple old anatomies?

A new comparative morphology has resulted from modern views of animal structure. The old comparative anatomy, in my early years, had become bogged down in the anatomy of **the** dogfish, **the** frog, **the** lizard, **the** pigeon, **the** rabbit. It had forgotten that there are many kinds of cartilaginous fishes, amphibians, reptiles, birds and mammals, respectively. This has now been corrected with new burgeonings of comparative morphology that look at diversity and complexity (e.g. again in date order: and starting with an older but very percipient text, Hyman, 1942; Hildebrand, 1974; Hildebrand et al., 1985; Cartmill and Smith, 1987; Stern 1988; Arthur, 1997; Oxnard 2008; Kardong, 2009). As a result, it provides new evidence of an underlying pattern for the anatomy of humans.

A new functional anatomy is resulting from advances in bio-engineering and bio-mathematics. It greatly modifies what could be estimated from the old anatomical inferences about function that were the main evidence presented in my earlier years. Now new concepts are emerging in testing and understanding the adaptations of anatomical structures (e.g. again, some quite a long time ago but developing up to the present: Stern and Oxnard, 1973; Wainright et al., 1976; Currey, 1984; Oxnard et al., 1990; Nigg and Herzog, 1994; Carter, 2001; Carter and Beaupré, 2001; Oxnard, 2008). Who, a few decades ago, would have guessed that the mechanics of fiberglass might illuminate the mechanics of bone?

A new neurobiology is resulting from major advances in studying the nervous system (e.g. Young, 1966; Jerison, 1973; Kandel et al., 2000; Striedter, 2005; Ramachanandran, 2011). Nowadays the brain can be seen both through the underlying molecular mechanisms for its development and through many of the new non-invasive imaging techniques that allow it to be seen during function. Most importantly, for our purposes, these new studies are revealing relationships between the anatomy of the body and the anatomy of the brain, the effects of integrations and communications, that were unguessed only two or three decades ago.

Finally, a new evolutionary biology, is, itself, evolving. It is a new, holistic, subject that integrates ideas from the aforementioned advances in development, comparison, function and integration as they culminate in evolution (e.g. Arthur, 1997; Page and Holmes, 1998; Hall, 1999; Mayr, 2001; Striedter, 2005; Cartmill and Smith, 2009. This, as Mayr has said, "Is Biology".

1.2   For Whom Is This System Useful?

The System of Anatomy in this book will therefore be (as are the other books in this series) of particular use to **students of biological anthropology, human biology,** and **human evolution**. Such students have no particular requirement for individual pieces of anatomical information until they need them for specific reports, term papers, study programs, research projects, grant proposals, dissertations and theses, and even for beginning post-doctoral research. In order, however, to use them in such contexts, these students can benefit from overviews of human structure. Such overviews were previously obtained, like those of an older generation of medical students, by relying upon the old, often hated, feats of memory, or by copying from little understood large anatomical tomes. How much better if they could be garnered through understanding the scientific underpinnings of human anatomy, underpinnings that render human anatomy a living, exciting science?

This book will likewise be of value to all those other students who require an overview of human structure but who also find that memorization alone is too difficult, or too boring, or both. These include:

students in biology who are specializing in **whole organisms**, **development, comparison, function,** and/or **evolution**, and for whom humans are merely a specific exemplar;

students in various health related professions who need a generalized understanding of the anatomy of their patients;

students in human movement sciences for whom a knowledge of specific anatomies is critical but whose understanding will be enhanced by this general approach;

students in bio-engineering and **medical engineering**, especially **biomechanics** (e.g. **orthopedics**) for whom, likewise, special anatomy is critical but who need to realize the science behind what they can divine from the anatomy texts they need to consult;

students in various life sciences who are working in disciplines that use microscopic and chemical methods, such as **micro-anatomy, ultrastructural anatomy, physiology, pathology, biochemistry,** and/or **molecular biology**, but whose knowledge in various aspects of human micro-structure needs to be seen in the context of human meso- and macro-structure.

This book will not be of much value to **medical students** under current medical educational regimes, where, truly, there is no room for the detail of any subject, and particularly where what little anatomy there is has to be closely related to specific clinical problems. It may, however, be of major value to **premedical students**. In the USA, where medicine has long been a post-baccalaureate degree, these are individuals who take premedical courses in undergraduate degrees as a hopeful preliminary to medical entrance. In the rest of the English speaking medical world, the majority of medical courses are five or six year undergraduate courses that used to hold a lot, probably far too much, anatomical content. Many medical courses are undergoing major reductions of medical anatomy during the course of conversion to four year graduate medical degrees following upon some other undergraduate degree. For such students, this is often a **premedical degree**.

As a result, even in this arena, there will be large numbers of students, far larger numbers than just those who will eventually enter medical school, who will want **premedical anatomy**. Such premedical anatomy will not be the specific but reduced medical-problem related anatomy of the new medical curricula, but the **science-based human anatomy** of a **general education** or **liberal arts undergraduate degree**.

Finally this book may be of value to those **current students in medical schools**, **current medical school teachers**, and **current medical practitioners** who, while realizing the necessary limitations of the restricted anatomy that they are or were presented with, may, for good intellectual reasons, want an understanding of **how** and **why**, and not just **where**, things are in the human body.

1.2.1   From Dissection to Science

Anatomy traditionally involves the cutting of the body, whether 'real' (i.e. from the cadaver), presented (i.e. in the prosection or model), or 'virtual' (i.e. in the computer). Personal dissection involves cutting and observing, usually from the outside in, from skin to bone. Prosected or modeled 'dissection' involves examining what someone else has cut or made. Computational dissection may also be carried out by moving from the outside in but, interestingly, it can also be done (for example in an@tomedia) in other ways, for example from the inside out, thus gradually clothing the bones, or from the center to the periphery following through the nervous or other system of the body, or through the bodily regions where systems may appear to be conflated, or through the pathways of development and growth, or even through examination of separate islands of related information such as the scattered endocrine glands of the body. Such approaches provide new and improved roadmaps of the body (especially important to the medical student and practitioner). However, without further exposition, they do not easily provide scientific understanding. In fact they often do the opposite, especially through their testing mechanisms, tending to present the learning of human anatomy as a process that I call 'the naming of the parts'. The 'naming of the parts' is often how anatomy is taught and, even more often, how it is learned.

Such approaches primarily seem to involve memorization of large numbers of small details. It is true that such memory can be reinforced through the use of acronyms, eponyms, and mnemonics. It can be especially strengthened by using atlases, diagrams, models, computer programs, and, most of all, examinations that emphasize the reproduction of facts and names. All of this can be useful for individuals who think they do not require understanding but who need the information. **But it is as boring as hell.**

It has well been documented how memorization without understanding (except when memorized at mother's knee) disappears so easily. It is indeed one of the reasons why many practicing professionals, especially some in the non-cutting specialities (remembering that hateful memorizing activity as students) hate anatomy. This reaction in some current medical professionals is partly why anatomy has been all but eliminated from the medical curriculum. Yet other practicing professionals recognize the enjoyment that they had in anatomy and the lifelong friendships they made there. It was the only time in the old medical curriculum where the same small group of students knew, helped, and worked with one-another and a single staff member for a whole year (even a whole two years in my ancient days!). In the new medical curricula this is now largely lost. In fact, no part of medical education today lasts long enough to provide this element of student and staff camaraderie.

My own experiences with science, medical, research and health professions students and practitioners, over more than 50 years, tell me that it can be different. Even today, there are still some medical and health practitioners who are interested in understanding the science underlying anatomy. They understand how anatomy provides them not only with the names of anatomy, but also with the vocabulary of medicine, the grammar and syntax of medicine, the ability to speak, write and understand the language of medicine, and help with unravelling the problems of medicine. They understand how communicating the science in medicine stems as powerfully from anatomy as from any other medical subject. They understand how knowledge of anatomy can be derivative, and how it can allow them to work out the anatomy, if forgotten, when needed. (My own general practitioner has anatomy books, atlases and models on his desk as tools to help him explain to **his** patients what is going on in **their** bodies).

1.3   What is This System?

The method that I have developed, over the years is, in contrast to 'the naming of the parts', an holistic integrative approach to human structure. It involves understanding that the structural details of the human body are the end results of a series of biological processes. These produce anatomical pattern due to:

change from differentiation and growth **over developmental time**
diversification through comparisons of different forms **living at the same time**
adaptation to mechanical and other factors **during functional time**
interaction between body and brain, brain and body, **through mind time**, and
innovation in structure resulting from evolution **during deep time.**

A number of important older texts did attempt to utilize these approaches but they were limited by what was known in those earlier days of the scientific disciplines of development, comparison, adaptation, integration, and evolution. Today, the excitement of totally new concepts in these disciplines cries out to be used to illuminate a new human anatomy.

This new way of approaching the anatomy of the body starts at the opposite end from the traditional. Instead of memorizing the fine detail of the body as dissected, prosected, or demonstrated as a geographical road-map, the principles of body construction are looked at from their main beginnings, whether those beginnings are developmental, comparative, adaptive, integrative, or evolutionary.

All this can be particularly illuminated by the new visualization techniques of the last few years. Thus the **blueprints** and **tool kits** of developmental change can be demonstrated, for example, through imaging of experimental chimeras wherein cells of different ancestry (quail in a chick, male in a female) can be separately defined and followed. The **plans** and **patterns** of comparative differences can be examined using mathematical methods for comparing structures available only to today's scientists. The **adaptations** and **optimizations** of function can be shown, for example, through engineering methods that demonstrate not only the structure, but also that structure when functioning. The **communications** and **controls** of the integrative brain can be illuminated by molecular dissection and three-dimensional non-invasive imaging. The **architectures** and **lifestyles** inherent in evolution can be revealed by combinations of all of the above.

These approaches to human anatomy: **developmental, comparative, functional, integrative** and **evolutionary;** and their internal mechanisms and processes: **blueprints** and **tool kits, plans** and **patterns, optimizations** and **adaptations, communications** and **controls,** and **architectures** and **lifestyles;** are all well known as separate expositions in their own disciplines.

There have been few attempts, however, to bring all of them together to provide the scientific underpinnings of a specifically human macroscopic anatomy. For example, most of the ideas above have been explicated in animal forms. There have been almost no attempts at all to integrate the new knowledge of recent years that, separately exciting in each of these disciplines, can now provide the holistic excitement that gives a truly scientific base to understanding the macroscopic anatomy of the specifically human body.

Yet it must be emphasized that this book **is** an exposition of the science underlying the macroscopic anatomy of the human body, and **not** of the separate disciplines of the developmental, comparative, functional, neural and evolutionary biologies themselves. These latter are the subjects of their own major disciplines, their own curricula, their own books. Yet macroscopic human anatomy is powerfully informed by both the old and new developments in these underlying sciences that are so basic to the understanding of biological structures in general.

1.4   Why, Therefore, This Book?

As a result, the approach that is taken in this book is quite different from that in the usual run of Human Anatomy texts today. Where, for example, the student, the teacher, or even the examiner, may ask, is the information about the triangles of the neck, the posterior abdominal wall, the axilla, and the branches of the maxillary artery, or even, the maze of arterial anastomoses around the human knee joint? The answer is that **they are not here**. They can already be found, in greater or lesser detail, in any large or small anatomy book. Anatomical information does appear here, of course, but only in the explication of pattern resulting from the various underlying sciences. It is provided, therefore, in a largely new guise.

This book, thus, does not mirror, nor is it meant to, any of the very large number of regional, systematic, dissection-based, clinically-based, or even summary human anatomy texts, whether large or small, exactly because those books do not adopt this special approach. Such books remain, nevertheless, most useful, indeed important, to students because they are sources for more detailed anatomy of specific regions when required.

Yet is it really the case that these new approaches have not been attempted before? In fact, following on Darwin's ideas, even before his time, many individuals tried to use scientific underpinnings as ways of understanding human structure. One of these was Todd and Bowman's Physiological Anatomy of 1845 at a time when it was clearly understood that anatomy and physiology should not be separated. Others were the understandings of vertebrate anatomy by Owen in 1868, and by Wiedersheim in 1882. This last was used to illuminate a specifically human anatomy (Wiedersheim, first edition in 1887). Yet another text that also unashamedly used this approach (so how new is it?) was Sir Arthur Keith's 'Human Embryology and Morphology' (Keith 1902). Of course, there have been enormous changes in physiology, embryology and comparative anatomy since those and other similar books. As a result new books have, over the years, really been needed. But I acknowledge that the idea is not really new – very few ideas are truly new – only the concepts to be integrated are new.

Among other anatomically based texts that do contain elements of the approaches suggested here are a number of books from later in the 20th century. They include, for example, Young's three far-sighted books: 'The Life of Vertebrates' (1950), 'The Life of Mammals' (1966) and 'An Introduction to the Study of Man' (1971). Likewise, but in a different vein, yet also employing elements of the approach implied here, is Hildebrand's wonderful 'Analysis of Vertebrate Structure' (1974). These, and others not cited, are excellent books, and still, in my opinion, very useful to students. But of course they were originally written many years ago and as a result do not include the mass of new developmental, comparative, functional, integrative and evolutionary information that has been elucidated in recent years. More recent are modern textbooks of vertebrate anatomy (e.g. Kardong, 2009). Most importantly, however, unlike the earlier essays by Wiedersheim and Keith, these books are not aimed at human anatomy, other than including humans as a minor example of another mammal, another tetrapod, another vertebrate, and so on. To be very fair, they were not aimed at understanding humans, specifically, in the first place.

An approach like that adopted here also exists in a number of excellent 'evo-devo' books published very recently. These, naturally, do include the new information. Indeed, these books, and the primary literature on which they are based, have been most important in deriving some of the underlying concepts for this book. They are excellent references for students wanting to take the combined developmental and evolutionary concepts further. They include, among many, for example, Hall's 'Evolutionary and Developmental Biology' (1999) and Twyman's 'Developmental Biology' (2001). Though these books include the evolutionary/developmental information underlying human anatomy, they, too, are primarily aimed at explicating the evolutionary and developmental anatomy of many other forms: primates, mammals, tetrapods, vertebrates, even invertebrates, and so on. They are excellent books. Their thrust, however, is not human anatomy specifically. The human anatomy is secondary, and, when it is included at all, is abbreviated, appropriately in context it must be said, as just another animal in the broader picture.

Yet another set of useful books include, among others, primary embryology texts. Larson's 'Human Embryology' (1993 and later editions), and Carlson's 'Human Embryology and Developmental Biology' (2004), figure among these. They are excellent texts, and they **are** aimed at the human situation. They are, however, specifically human embryology texts. They do provide many important and fascinating linking materials from other species to humans, but they remain treatises separate from human anatomy *per se*.

The final set of books that use elements of the approach adopted here are genuinely human anatomy books. They include Cartmill, Hylander and Shafland's 'Human Structure' (1987), and Stern's 'Essentials of Gross Anatomy' (1988) among a number of others. The former of these two books (Cartmill, et al.) provides much explanatory information from developmental, functional and evolutionary biology. However, though it gives short useful introductions to these matters, because it was written some time ago, it does not include some of the more recent linking material from other disciplines. Further, its organization is still largely based on the traditional anatomical regions. Its primary aim is to fully cover the anatomy of the human body at an appropriate level, with the result that the developmental, functional and evolutionary information is integrated into the human anatomy in only a secondary manner.

The second of these two books (Stern) is an even more standard human anatomy text for medical students. Yet it, too, integrates considerable information about developmental, comparative and functional aspects of human anatomy. For example, in one place, Stern gives an excellent description of the changes in limb form during limb evolution, common in vertebrate anatomy texts, unusual in human anatomy texts. In a different vein, however, he gives a mnemonic for remembering aspects of knee anatomy: "The professors who taught me my anatomy ... " he says were " ... [the late] Ronald Singer and Charles Oxnard. Their initials – RSCO – remind me that if I could peer through my femur down onto the top of the tibial plateau of my **R**ight **S**ide, I would see the letters **C** and **O** formed by the medial and lateral menisci, respectively ... " Stern notes that "The value of this mnemonic to persons not trained by Singer and Oxnard is unclear".

I am proud to have figured in Jack's description and regret that Ron Singer is no longer with us. As with Jack's codicil, however, I cannot believe that his mnemonic is a good scientific reason for understanding the shapes of the menisci! The functional explanations related to load bearing and movement of the femur on the tibia during knee function are so much better! The general arrangement of Stern's book is, again, and appropriately so, quite classical. The conceptual parts are more in the form of useful and important *obiter dicta* in a mainly human macroscopic anatomical text.

Finally, I draw your attention to a beautiful book by Shubin, also from Chicago. This book: "Your Inner Fish" is truly a journey into the 3.5 billion year history of the human body. It should be read by all who might find the present book useful. Again, however, it does not replicate this book. Its emphasis is truly on vignettes on "the long long journey" to human anatomy, rather than on the final effects on human anatomy itself. It is, nevertheless a wonderful introduction to the story being told here.

Let me emphasise that I am not criticizing any of these texts as being inadequate, quite the reverse. I owe much to them and to many others not cited. In fact, all these other books would be most valuable when used alongside the current text. I am, however, trying to emphasize differences from the approach adopted here.

1.5   What is My Hope for My Readers?

One hope is that this book will introduce the student to the use of the scientific underpinnings of human anatomy as the major rationale for understanding it.

A second is that the student will realize that the book makes no attempt to cover everything in anatomy in an equal manner. It certainly does not try to capture the whole of human anatomy. As an example, the book goes into considerable detail about the combined developmental and comparative underpinnings of human variation in the proximal parts of both limbs, but it does not address the equivalent distal limb parts in anywhere near so much detail. Likewise, other regions of the body are covered in this patchwork manner.

Why?

One reason is the intention to introduce students to principles in some parts, and to add to their learning by leaving it to them to work out the extensions of the principles in other parts. In other words, I want the students to build on the partial story that I am providing.

A second reason, however, is because the principles have not yet been equally and fully worked out in all areas. I myself have not yet worked in all areas. For example, much more is known about the head than about the trunk, much more about muscles, joints and bones, than about tendons, ligaments and fascial sheets. As a result, some of the concepts that I suggest here may well turn out to be wrong! And other concepts, apparently equivocal or even unlikely, may well turn out to have been right. In other words, I want the students to see that I don't know the full story, that possibly at this time no-one does, and that, therefore, there are more questions to be asked.

A third reason is that the complete understanding and then integration of all these basic sciences and their application to human anatomy is too much to ask of any single author, certainly of this author. In other words, this is only a work in progress.

A final reason is the hope that through these partial answers, some right and some wrong, students will come to realize the many other interesting questions that they can ask in their own researches, whether expressed as simple curiosity, term papers, undergraduate honours projects, practical applications, postgraduate theses and dissertations, or even later, academic studies, papers and books, as the students, in turn, progress. In other words, I want them to be stirred into undertaking their own works in their own futures.

In this context I have never forgotten the words of one young academic when I was asked to review the status of a particular anatomy department. This was a young man, absolutely fascinated by anatomy, outgoing to his students, deeply concerned with teaching as well as he could, and I could see, a teacher very much liked by his students.

"What about your research?" I asked at one point in the interview. He told me.

Later at one of the social occasions that occur in such situations, I met him again over drinks and repeated my question.

"Tell me more about your research in human anatomy."

"Well" he said, "to tell you the truth I am not very interested in my research. I do it because I know I have to publish for my career. I do it because it gives me something to tell people like you when you come around."

"So, why are you so interested in teaching human anatomy?" I asked.

"Because I love the structure of human anatomy" he said.

"Because I love how complete it is".

"Because I love to present this completed structure to the students".

"*I love the fact that it is so complete*" he continued, "*that there are no more questions to be asked*".

No matter how good and attractive this teacher is to his students, if he leaves them with the notion that human anatomy is complete, that there are no more questions to be asked, then I aver that he is a very bad teacher indeed.

It is my fervent desire that this book should show how the new work in the other disciplines underlying human anatomy can provide new questions for anatomists.

It is my fervent desire, too, that these views of anatomy, will present new questions for these other disciplines.

It is my final fervent desire that new questions will come thick and fast to the reader.

CHAPTER TWO

A Bird's-eye View of the Human Body

2.1   The Scientific Basis of Anatomy

There has always been a scientific basis to learning human anatomy, even if this existed in the minds of only some teachers and investigators, and quite rarely indeed in the minds of students. The main way of teaching and learning human anatomy was not really science, but a gigantic task of memory, of lists of structures, of tables of relationships, of maps of the body, of origins and insertions of muscles, of artificial gateways to regions, even of mnemonics, and so on. These were, and still are, very useful aids that have passed many students through their examinations. In recent years, however, there have been major changes in the sciences underlying human

---

The Scientific Bases of Human Anatomy, First Edition. Charles Oxnard.
© 2015 John Wiley & Sons, Inc. Published 2015 by John Wiley & Sons, Inc.

anatomy. These changes have been extensive enough that they can now become the primary components for understanding human structure. They have transformed the original older disciplines of embryology, comparative anatomy, functional morphology, neuroanatomy and evolution.

2.1.1 From Embryology to a New Developmental Biology

Thus, a new developmental biology (e.g. Hall, 1999) has arisen from classical embryology, resulting from deeper knowledge of genes and other molecules responsible for developmental mechanisms and products in humans. This is showing us a new GUT (grand underlying theory) of anatomical structure almost undreamt of even only a few decades ago. There was a time when there was a great gulf fixed, for example, between creatures without backbones and animals with backbones, between compound eyes in fruit flies and human eyes, between chicken 'teeth' and human teeth, between whale brains and human brains (e.g. Shubin, 2008). Who, in those days, would have realized that similar genetic mechanisms and underlying molecular processes would be found common to each? Who would have thought that the mechanism for producing a limb could act anywhere within the body (Fig. 2.1(a,b)).

2.1.2 From Vertebrate Anatomy to a New Comparative Morphology

A new comparative morphology (e.g. Hildebrand et al., 1985, through Kardong, 2009) has arisen from the older vertebrate anatomy (e.g. Owen, 1866) that is resulting in a clearer understanding of the place of human structure amongst animal structures. Many scientists long understood something of the implications for humans of the comparison of the forms of different living animals. Some even applied it to understanding human structure, though usually only as minor asides. The converse was that, though it was well understood by zoologists, it was only applied to human structure in perhaps the last few paragraphs of major volumes on vertebrate anatomy.

In particular, much of the knowledge of comparative anatomy that can be applied to humans stems from understanding the anomalous features (often called variations) of humans. Indeed, for most of the last century, such studies were perceived as little more than anatomical stamp-collecting, useful in applied medical and especially surgical situations, but certainly scarcely hard science. Yet today they may be of supreme importance (Fig. 2.2 (a–d)). The comparative anatomists of prior centuries knew a lot about the importance of human variations even though they operated without any knowledge of genetics and with only little information about development. How many, even of these workers, could have realized how all-enlightening such comparisons would become? How far from 'stamp-collecting' they would travel, as evidence of common underlying developmental patterns?

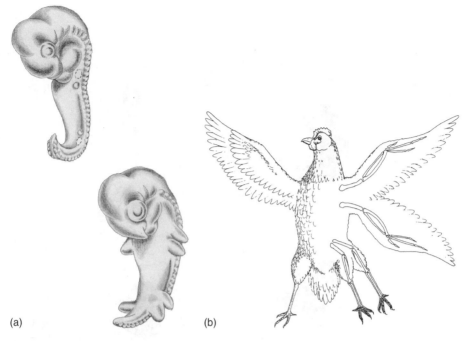

(a)   (b)

FIGURE 2.1

The generation of an extra pair of limbs on one side by experimental manipulation in (a) bird embryos and (b) bird adults (b).

2.1.3   From Functional Anatomy to a New Adaptive Morphology

Third, a new functional morphology resulting from experimental applications of advances in engineering is transforming the inferences of an older functional anatomy by testing them in new ways. This has produced new concepts in understanding the adaptation of anatomical structures to function. Who, a few decades ago, would have guessed that mechanical phenomena in fiberglass might resemble those in bone; that elasticity of rubber would have applications to the functions of tendons, that the forms of trees and the branching of rivers could provide ideas enlightening the patterns of arteries and bronchi (though D'Arcy Thompson as long ago as 1917 was aware of some of this)? Studies of the engineering of bone (Fig. 2.3 (a–e)) and bones (Fig. 2.4) show the promise of increasing our understanding of specifically human anatomy (e.g. see Oxnard, 2008).

2.1.4   From Neuroanatomy to a New Neurobiology

Fourth, a new neurobiology (e.g. Kandel, et al., 2000, Strieter, 2005, Ramachanandran, 2011, and many web sources on brains and brain functions) has resulted from major new methods of studying the nervous system. These now run the gamut

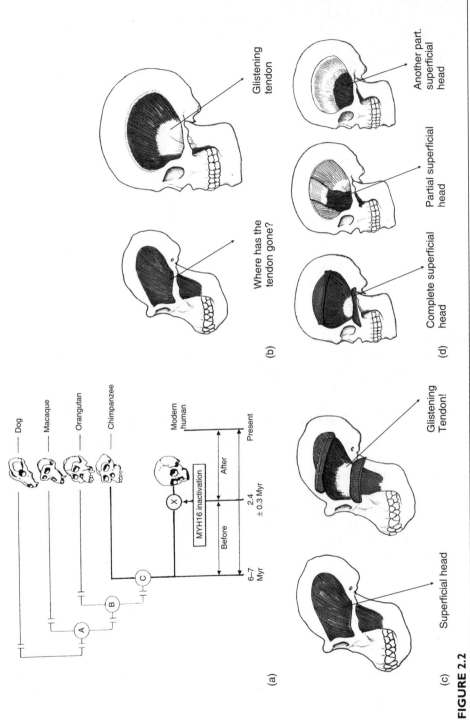

FIGURE 2.2

(a) A phylogenetic view of the molecular system, MYH16, responsible for the reduction in humans alone among primates, of the muscles of mastication, (b) the difference between a normal human and a chimpanzee, (c) the fact that the chimpanzee has a superficial layer of the temporalis muscle that is absent in humans, and (d) the fact that a few humans do still have from gross anatomical and low power microscopic recognition of individual parts of the brain, through standard microscopy (as it was when I was a medical student), through ultrastructural microscopy, through cellular and sub-cellular events, to ultimate brain molecules.

Especially important have been functional investigations that have progressed from studies of interference with function by crude macroscopic brain lesions, through studies of carefully placed micro-lesions, through intracellular recordings, and through interactions along brain cell axons and across individual synapses between cells, to underlying chemical events associated with function.

Developmental mechanisms in the brain have come to be understood at finer and finer levels, from overall organogenesis, through mechanisms of movement of cell layers, through growing populations of cells and cell deaths, to the genetic basis of brain development and the many complex molecular factors that govern it.

Perhaps most of all, these apparently separate approaches: structural, functional, and developmental have started to coalesce so that nowadays the brain can be seen developing and functioning through many of the new non-invasive imaging techniques. All this allows a better understanding not only of brain structure but also brain function providing whole new views about this most central of organs in humans (Fig. 2.5). Also important for our purposes, however, these new studies are also revealing integrative relationships between the anatomy of the human brain and the anatomy of the human body, of controlling mechanisms, that were unguessed at even only two or three decades ago.

2.1.5   A New Evolutionary Biology

Finally, as a result of all these, a new evolutionary biology is itself, evolving. This presents ideas that are now so much more clearly revealed through the union of the aforementioned advances in molecular development, comparative morphology, functional adaptation, and brain linkages, all culminating in the evolutionary process itself. They involve a clearer understanding of how to describe evolution and how the analysis of information from animals tests ideas about human structures (e.g. Page and Holmes, 1998; Hall, 1999; Carey, 2012). All this integrates ideas from the aforementioned advances in development, comparison, function and the brain culminating in the evolutionary process, which, as Mayr (2001) has said "Is Biology". Even so, it must be recorded that there is still a great deal more to discover (Fig. 2.6 (a–d)).

2.1.6   Implications: New Impacts on Human Anatomy

As a result of all these changes and their interactions, it is now possible to understand human structure in a new way. This is a way that melds with current concepts of understanding animal structure (as already well-described by vertebrate anatomists). Thus, in contrast to the 'naming of parts' method of learning human

Change in pattern when diagonal loads remodel the architecture

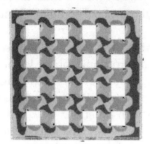

Beginning structure

After three iterations

After six iterations

(a)

FIGURE 2.3

(a) The changes that can be induced in a model of a right angle (orthogonal) network of bony spicules (mimicking bone trabeculae) when they are subjected to forces at an angle different to orthogonal. The trabeculae gradually change in orientation until they become aligned with the new set of orthogonal forces. (b) The changes that can be induced in an orthogonal network when acted upon by loads greater in the vertical than the horizontal direction. The vertical trabeculae become thick, the horizontal ones become thin. (c) The changes that can be induced in an orthogonal network when a microfracture is simulated. (d) The patterns of trabeculae showing thicker vertical than thinner horizontal trabeculae in a section of a real vertebra in a person afflicted with severe osteoporosis compared with normal. The similarity of (b) and (d) is clear. (e) The worst case of osteoporosis (in a long term quadriplegic) that I have ever seen.

Change in pattern when diagonal loads remodel the architecture

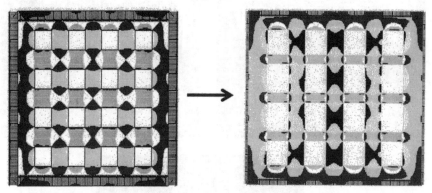

Beginning: equal cross structure   End: unequal thick and thin structure

(b)

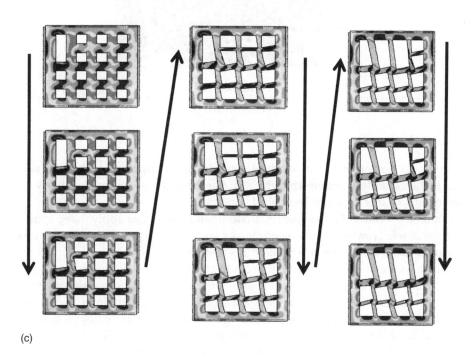

(c)

FIGURE 2.3

(Continued)

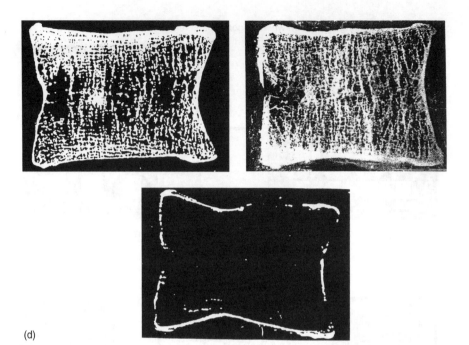

(d)

FIGURE 2.3

(*Continued*)

anatomy so prevalent in our medical schools, is an approach to human structure that involves understanding that the structural details of the human body are the end results of this series of biological concepts.

A number of older texts of human anatomy did attempt to apply these approaches as exemplars. However, most such books were limited by what little was known in earlier days of the above scientific disciplines, and also by a perceived need to use most of the effort and space to provide all the details of human structure in the traditional way. Because of the descriptive excellence of these older texts it is no longer necessary to repeat all the detailed anatomical information in a book like this. Today, however, the excitement of totally new concepts in these disciplines cries out to be presented to illuminate human anatomy in a new way. This has already been well achieved by new textbooks applied to animal anatomies but has scarcely been applied to human anatomies.

This new way of approaching the anatomy of the body starts at the opposite end from the traditional. Instead of memorizing the details of the body as dissected, prosected, demonstrated, or even studied computationally in a geographic manner, the principles of body construction can be looked at from its beginnings, whether those beginnings are developmental, adaptive, comparative, integrative, or evolutionary.

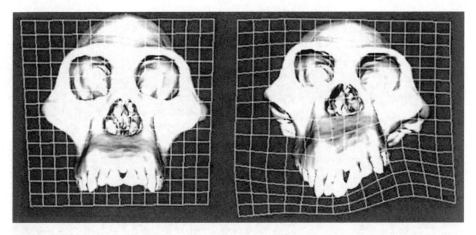

FIGURE 2.4

Mathematical strain analysis of a monkey skull, showing, during a bite, how the skull is deformed. The deformation is multiplied many times to allow the reader to see it.

What shipwright would try to understand a ship by taking it to pieces, rivet by rivet? Shipwrights know how a ship is built starting from plans (development). They improve a ship by assessing what features of the ship worked last time (adaptation). They understand the differences in ships built for different purposes (comparison). They realize the importance of communications mechanisms in ships, so enormously elaborated in these days of electronics and computers (integration). They recognize the changes in ship-building techniques that have occurred over the generations of ship-builders (from archives); that is, through evolution, though not, in the case of ships, biological evolution. These are most useful bases from which to understand the structure of ships and the way they work. So too for the body.

Nevertheless, this does not mean that dissections and the details are not important. An archeological shipwright may need to take the remains of a 'fossil' ship to pieces for comparison with the 'anatomies' of 'living' ships in order to understand novel or unexpected ways in which the shipwrights of the past thought and worked. They might even then rebuild 'fossil' ships from the extant fragments that they find. An anthropologist may likewise need to use pieces of fossils, for comparison with the anatomies of living humans and their relatives. And reconstruction is a *sine qua non* for the anatomist and the anthropologist. Except that we often get it so wrong. Certainly, most of today's medical students are no longer required to take the human body to pieces, though even this is starting to change back. Some medical schools that largely left anatomy and dissecting are returning to them!

These ideas mean that we can start with what may seem to be a small number of very simple processes and structures in the early embryo.

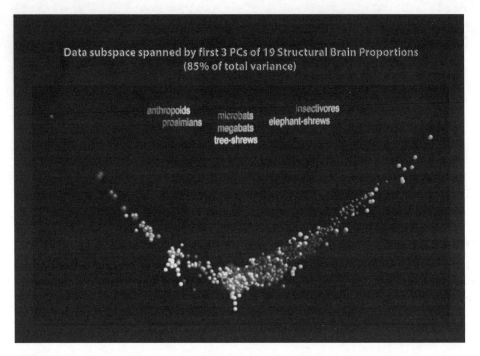

FIGURE 2.5

Analysis of brain sizes separates some different species of mammals. Primates (red and yellow dots diagonally upwards to the left) are organized completely differently from insectivores (various shades of green diagonally upwards to the right) and bats (various shades of blue horizontally organized at the base of the diagram). (*See insert for color representation of this figure.*)

2.1.7 Looking Upwards Through Ontogenetic Time

We can, thus, look upwards from the fertilized egg to the complexities of final adult form. This is the developmental approach to the structure of the human body. Textbooks of embryology and developmental biology usually spend large amounts of time on the earliest developmental processes. They sometimes spend much less time on what happens later. But a great deal of what happens later provides great understanding of human anatomy.

Developmental biology in itself is a fascinating subject that has truly burgeoned in recent years. In human anatomy, however, it is almost always taught as a separate subject in textbooks of embryology. Conversely, many anatomy texts, though they do throw in occasional examples from development to help enlighten difficult areas or apparently unusual relationships, do not usually use development as the major plan for understanding adult structure. Yet there have been many new findings in today's developmental biology that have applications to human anatomy.

These include information about the genetic determination of structures, about the cascades of post-genetic molecular processes modifying the effects of not only

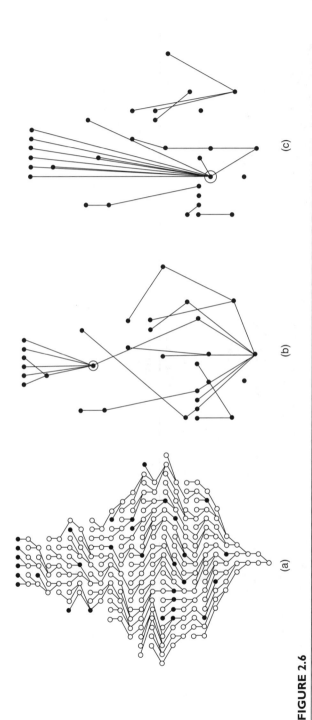

(a)

(b)

(c)

FIGURE 2.6

A model (a) of all the species generated computationally from the splitting (or not) of species over time. The single species (black circle) at the base of the diagram is the starting species. The black circles at the top of the diagram are the current living species. Open circles throughout the tree are the many species that were never found as fossils. Black circles within the tree are the few species found as fossils. The true relationships of all species are shown by the tree of lines joining them. This tree of relationships can be compared by the tree of relationships (b) that is found to lie 13 generations back from the present day. This is further exemplified by a different model (c) derived from the data from only the fossil species (i.e. the usual paleontological technique) to find the common ancestor of all the living forms. That common ancestral species lies, in this example, only five species generations back from the present day. This can be compared with the real common ancestral species (d) containing a bottle neck. The common ancestral species as defined by the fossils alone lies only at about the level of the bottle neck. The real common ancestral species is many generation further back and results from a number of different species lineages (as shown by the lines in red, green and orange). In other words, looking at fossil data is rather likely to give incorrect species lineages, and incorrect times by factors of 3 or more. (*See insert for color representation of this figure.*)

23

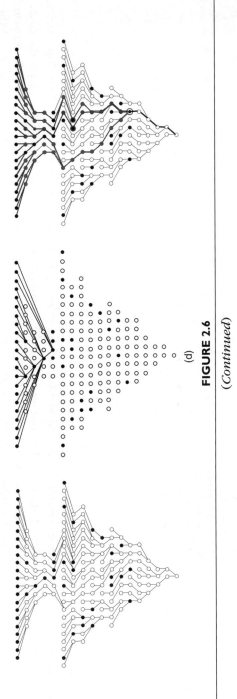

(d)

FIGURE 2.6

(*Continued*)

particular genes but more especially of clusters of gene sequences, and about the impacts, upon these mechanisms, of a wide range of environmental factors. They involve mechanisms and processes that result in cell proliferations and movements, of cell deaths and cellular communications, and of changes in populations of cells producing patterns, forms and structures. They especially involve the times at which things occur, including differences through the same mechanisms producing different structures when they operate at different times. They are influenced by external, non-genetic, factors that modulate how internal genetic and molecular factors display themselves. All of these new developments allow insights today into detailed adult structure as never before. They inform us about the 'blueprints' that underlie human structures, the 'tool kits' that produce them, and the 'clocks' that influence them.

2.1.8 Looking Inwards Across Comparative Time

We can also start with the structural similarities evident in a series of animal species all living in the present, and, through such comparisons, look inwards towards the special structural patterns specific to humans.

This is the comparative approach to the structure of the human body. In earlier days, it was traditional in textbooks of vertebrate anatomy to spend large amounts of time on the structures of other species for their own sake. Such texts, even if they did not ignore applications to the understanding of human structure, nevertheless resigned themselves to a few cursory examples. Likewise, human anatomical texts of earlier days, though they did not totally ignore comparisons with other species, usually only entered them as occasional 'explanations' of one or another apparent human anatomical idiosyncrasy (as it was thought).

Yet new 'whole-organism' morphology, based in genetics and development, affected by environment, and reflective of underlying evolutionary mechanisms, now greatly informs on the common principles, the underlying patterns, girding vertebrate structure. It seems such a small step to use these in the elucidation of human structure. They indicate something of the underlying 'body plans' descriptive of the features common to different animal organizations. They are an incomplete 'archive' of structural remnants remaining over the span of evolution. They involve 'modulation' of features characterizing separate species today that especially illuminate the human body.

2.1.9 Looking Along Functional Time

This then leads onto the examination of the adaptive functions which the detailed anatomy of the human body subserves: the functional approach to understanding human structure. Again it was traditional that the older textbooks of anatomy went into only a little detail of how the anatomy works. This was partly the result of a two-centuries-long boundary dispute between anatomy and physiology. It was also partly because it was considered less 'scientific' to 'speculate' about function.

The very word, 'speculate', implied that, somehow, it was pejorative. Knowledge about function was, thus, not usually placed within the contexts of anatomy. When these texts did go into function, they usually did it through anatomical inference, an estimation of what function the anatomy could allow, not a discovery of what it actually sustained. (Nevertheless, these older writers were very well aware of functional implications; they just did not write about them!)

But a new functional morphology has evolved that is resulting in a far greater understanding of the 'mechanics' of anatomical structures. It recognizes 'optimization' in anatomical structures. It is shaped by 'limiting factors' of function. It can 'integrate' many different aspects. Though mostly worked out in other species, it too can be applied specifically in the understanding of human structural adaptations.

2.1.10 Looking Through Integrative or Control Time

Further, this allows us to look at the way in which body structure is related to brain structure, the way in which body function interacts with brain function. Thus, although the structure and function of the brain, indeed of the whole nervous system was originally a part of human anatomy, the more that that was learnt about the nervous system the more it was separated off as a separate specialty.

This similarly happened to function – physiology became separated from anatomy when the textbooks of anatomy and physiology and the Society of Anatomy and Physiology, separated into their separate books and societies in the 19th century. (Interestingly, in the new premedical courses and text books in premedical undergraduate education in the United States, anatomy and physiology are being put back together again).

This separation was then followed by the further separations of microanatomy, histology, from meso-anatomy (what can be seen with a hand-lens), and, in turn, from macro-anatomy (what can be seen with the knife and naked eye). Further, cellular biology, ultrastructural biology, chemical biology, and so on, have all become separate from the original human anatomy.

Of course, these separations and specializations were extremely important for the development and growth of these disciplines, witness the enormous advances that have resulted from them. But in the mind of the student this often meant that human anatomy was seen merely as the map of the structures. The teacher, in contrast, usually knew the relationships between these disciplines and basic human structures. But in most recent times even the teacher (whose own scholarly work is likely in some discipline apparently far removed from human anatomy) sometimes views the teaching of human anatomy as merely a matter of presenting the geography of the human body for applications in clinical medical problems.

It is only now, when the integrative or holistic view of human structure is gaining precedence, that the relationship of brain structure to body structure once more becomes important. The patterns in body anatomy are so strongly related to the new patterns in brain anatomy that are being worked out. Study of the brain is not just its own subject, but a subject integrating the whole of human anatomy.

2.1.11   Looking Forward Through Evolutionary Time, Through Deep Time

Looking forward through deep time, the phylogenetic approach, combines information from the differences in molecular and developmental processes, the comparisons of species today, the workings of anatomies, and the integrations of the brain, and applies them to the changes from species in the past to provide an evolutionary picture. Though this last approach combines all of the others, it suffers from deficiencies. Most of the species of prior times are not fossilized. Even the fossils we have usually only provide inferences from bones. Even the bones are largely fragmentary or distorted. Nevertheless, this overall approach is important in helping understand the complexities of the human body.

2.1.12   New Visualization Techniques

Each of these approaches can be especially illuminated by the new visualization techniques of the last few decades. Thus, for example, genetic and developmental changes can be demonstrated through imaging of chimeras wherein cells of different ancestry (e.g., quail in a chick, male in a female) can be separately visualized. Comparative differences, for example, can be revealed by analytical imaging techniques that involve subtraction (e.g. the shape of a brain endocast shown in the computed tomographic scan (CAT scan) of an otherwise solid fossil skull. Or they can be demonstrated by comparisons of differences in shape (e.g. the differences between shoulder blades in animals of different lifestyles as seen through the use of thin plate splines).

Further, imaging methods may demonstrate not only the form of a structure but also the form of that structure when functioning. Perhaps the most well known example is the brain, but, of course, many other organs, for example, blood vessels, can also be so illuminated. Integration and control of structure can be related to advances in brain structure, function and development.

Imaging methods can also be used to test evolutionary hypotheses using new molecular data (after all the molecules are likely to be somewhat closer to evolution than the structures that result from them). Molecular data are now flooding in: see the comparisons of Neanderthal and Modern DNAs – and these imply that Neanderthals are rather less different from us than many have thought. Such hypotheses can also be tested using mathematical methods to look at change and difference in shape and form, engineering techniques to look at function of form and architecture, and modeling methods to look at evolution of lineages of individuals and separations of species. All this is explicated through the concatenation of knowledge of development, comparison, function and the brain.

These approaches to anatomy are all fairly well known as separate expositions. Certainly all have been applied to the understanding of the vertebrate anatomies as a whole. There have been few attempts, however, to apply them to specifically human anatomy. And there have been almost no attempts to integrate them, exciting in each of these disciplines separately, into a holistic view that promises to give a truly scientific base to understanding the human body.

It must be emphasized, however, that this book is an exposition of the anatomy of the human body, not of the other species that are frequently mentioned, and not at all of the separate underlying disciplines themselves. All these latter are to be found elsewhere in excellent texts. Yet the understanding of human structure is powerfully informed by the new developments in these other sciences.

2.2 Foundations: From Cell to Embryo

The foundations of human anatomy are to be found in the transition from the fertilized egg to the embryo. How this transition occurs and what is produced is fundamental to the subsequent anatomy.

2.2.1 'Blueprints' and 'Tool Kits', 'Clocks' and 'Environment'

The fertilized egg possesses a 'blueprint' of genetic instructions, and this controls a 'tool kit' of molecular factors carrying out and modifying the blueprint. The blueprint determines the production of the specifically human body. The tool kit manufactures the regular spatial arrangements of embryonic organization, such as segmentation, produces materials that lead to such tissues as muscle and bone, and constructs organs such as the heart and eye. These manufacturing processes are brought about by cascades of molecular factors, together with forward and backward influences along them. These, in turn, are operated on by 'clocks' and by 'environment'. The clocks alter what is produced at different times, and the environment interacts to change developmental expressions.

Let us start the process of understanding the anatomy of the human body by considering these mechanisms. The first stage is the production, from the first cell (the fertilized egg), of the comma-shaped embryo. Though more detailed descriptions are provided in later sections, a bird's-eye view of the process is useful as a beginning.

The initial blue print, is, of course, present in the fertilized egg, and, indeed, in every subsequent cell formed during the first few weeks after conception: the pre-embryonic period. The union of the sperm and egg produces a single cell (the zygote). The zygote divides several times and becomes, in turn, a ball of cells (the morula). As more cells are produced the morula undergoes 'cavitation'; that is, it becomes a hollowed-out ball of cells (the blastocyst). A plate of cells (the embryonic plate) then develops lying across the cavity of the blastula (now called the gastrula). Only this embryonic plate is the future individual. All other structures surrounding the embryonic plate are either involved in the production of the placenta and other supporting tissues that are not part of the new individual, or actually disappear at some stage during development. A great deal is known about the formation of the embryonic plate but we will ignore these further details as they are not necessary to see the big picture.

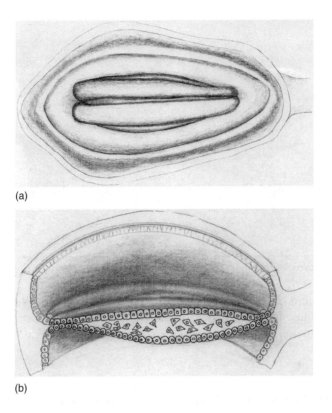

(a)

(b)

FIGURE 2.7

En face diagrammatic view of an early embryo (a) and its diagrammatic longitudinal section (b) showing the three germ layers, above ectoderm, below, endoderm and centrally, mesoderm.

2.2.2 The Two-dimensional Embryo: Determination of Right and eft

The embryo (embryonic plate) thus at first consists of a single layer of cells. It then becomes three layers of cells (Fig. 2.7(a, b) by processes that are now quite well understood but that, again, we will ignore here. These are often known as germ layers: **ectoderm**, **mesoderm** and **endoderm**. They lie uppermost, intermediately and lowermost, respectively, in the plate and already define what will be the major elements of the adult. In this flat oval plate, right and left halves of the embryo are quickly recognizable through the production of longitudinal grooves in the upper cell layer (ectoderm), and longitudinal differentiations in the middle cell layer (mesoderm). The central ectodermal groove is called, the neural groove. The central mesodermal structure is called the **notochord**. These are aligned along the long axis of the oval embryonic plate. They define the left and right halves of what is now a bilaterally symmetrical flat three cell layered structure (Fig. 2.8).

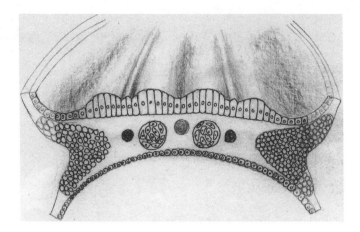

FIGURE 2.8

Diagrammatic cross-section of a slightly later embryo showing progression from Fig. 2:7 (a and b) of, above, ectoderm and neural grooves, below, simple endoderm, and between them, lateral plate mesoderm, intermediate mesoderm and paraxial mesoderm, this last lying alongside the axially placed notochord. Note: this figure Fig. 2.9 are 'exploded' so that the various structures are separated for a clearer are understanding.

2.2.3 From the Two-dimensional Plate to the Three-dimensional Sausage

The flat embryonic plate then begins to fold downwards on each side of the notochord and these folds eventually join below in the midline (Fig. 2.9 (a–e)). Because of the folding, what was the upper layer (ectoderm) of the embryonic disc now becomes an outer tube covering the exterior of what is now an embryonic rod. The main part of this outer tube will form, among other things, the epithelial part of the skin covering the adult. The aforementioned longitudinal groove will become, by a process that will be described later, the nervous system (hence, now, **neural tube**). At the junction of the cutaneous and neural parts of the ectoderm, is a special region called the **neural crest**. This forms a large number of different structures in the adult.

What was the lower layer (endoderm) of the disc likewise comes to form a tube, but one lying inside the rod enclosing an internal cavity. This internal tube becomes, among other things, the epithelium lining the **gut** of the individual. The space within it will become the **alimentary canal**.

The middle layer (mesoderm), lying sandwiched between the other two, provides, once the folding has occurred, a degree of solidity to the sausage-shaped embryo. The parts of the mesoderm closer to the ectoderm will largely become structures of the external body wall, forming the most superficial connective tissues (**dermis**) that support the epithelium of the skin, the intermediate voluntary muscles of the body wall, and the deeper centrally located supporting tissues (bones and other

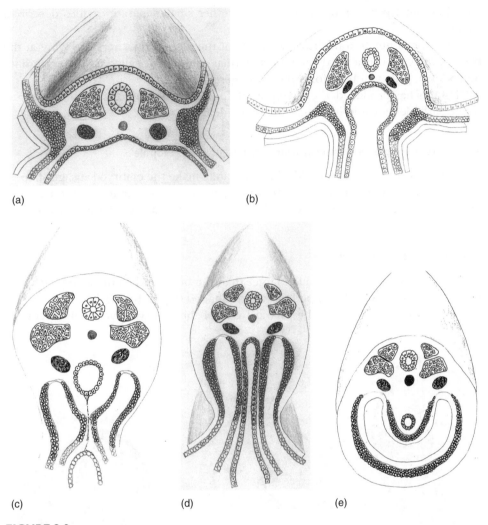

(a)

(b)

(c)

(d)

(e)

FIGURE 2.9

Progressions in the diagrammatic cross-section of Fig. 2.8 showing changes in positions of the various structures over time (a, b, c and d) resulting, eventually, in a completely closed body with gut placed internally (e).

connective tissues) of the body wall. The parts of the mesodermal layer nearer the endoderm will become some of the connective tissues and involuntary muscles that support the epithelium of the gut, together with the connective tissues that come to support various gut derivatives and organs.

Between these two mesodermal components develops a new space that forms a cavity within the body. This single cavity eventually becomes divided into thoracic and abdominal parts (and also one or two other small separate cavities described later).

At the dorsal junction of these two parts of the mesoderm, therefore lying at the back of the cavity, is a third small mesodermal component called the **intermediate mesoderm**. This will help form the urino-genital system.

All of this has produced a relatively solid three-dimensional sausage-shaped structure clearly demarcated into right and left sides.

2.2.4 From Simple to Complex, from a 'Sausage' to a 'Tadpole'

Concomitant with the lateral folds that ultimately make the embryo sausage-shaped, are other bends, folds, and differential enlargements that occur at its two ends. Each end of the bent sausage folds ventrally (i.e. away from the ectodermal covering).

Determination of Head and Tail

One of these folds becomes much larger than the other and so the sausage-shaped embryo becomes tadpole-shaped or comma-shaped (Fig. 2.10 (a)). The swollen part of the tadpole or comma is the forward end of the embryo, eventually becoming

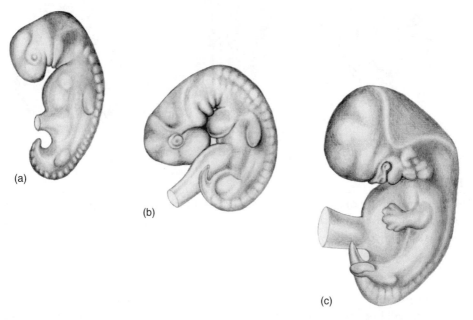

(a)

(b)

(c)

FIGURE 2.10

External form of the early embryo (a) with small protrusion for limbs, through a later stage (b) with heart and limbs more evident, to an even later stage (c) that is starting to resemble a baby.

the adult head (hence cranial). The tail of the tadpole, point of the comma, remains quite small and becomes the tail of the embryo (hence caudal) though of course there is no external tail in postnatal humans.

Determination of Back (Dorsal) and Front (Ventral)

The shallow neural groove on the now convex surface (ectoderm) of the comma deepens. Its lateral lips come together, and it becomes a tube that sinks beneath the surface. It forms the central nervous system lying below the future back (dorsum) of the individual (so it is dorsal in position). The concave surface of the comma, still connected to extra-embryonic structures, fuses in the midline as the future belly (venter) of the individual (ventral in position). (In human anatomy the terms anterior and posterior, meaning front and back are often used, but these are only directional words and lead to some confusion until the whole process is understood (see later).

The Next Stage: Building a Trunk and a Head

Within the three-dimensional comma-shaped embryo, the changes already described are evident as preparations for the formation of a head and trunk. Thus, the enlarged end of the comma starts to become the future head (Fig. 2.10 (b)). The remainder of the comma becomes the future trunk. These changes are not just matters of description (head versus trunk, **cranial** versus **caudal**) but describe a fundamental (yet complex as we shall see) structural differentiation within the embryo (Fig. 2.10 (c)). The head/trunk separation is initiated by molecular factors in the pre-embryonic period.

The Trunk

Although the head starts to differentiate earlier, the trunk is described here first because its formation is somewhat simpler than that of the head. In the trunk, two sets of structures are formed.

One of these comes from the actions of a gene series (homeobox genes) involving the notochord and the neural tube. They affect the ectodermal derivatives (skin, neural crest and nervous system) and the superficial parts of the mesoderm (lying originally alongside the body axis in the embryonic disc, hence paraxial mesoderm). This mesoderm is also called **somatic** (Greek soma, of the body) mesoderm and, of course, is overlain by ectoderm. These particular mesodermal and ectodermal layers together define most of what we can call the **external trunk**. They become associated with the functions of conscious sensations and voluntary movements.

The second set of structures stems from the actions of a homeobox gene series involving both the endodermal tube (eventually the gut) and the deeper parts of the mesoderm (originally lying laterally when the embryo was a disc, hence, lateral plate mesoderm). This mesoderm is also called visceral mesoderm because, after folding, it comes to lie alongside the endodermal gut tube (viscus: internal organ, plural viscera). Together these structures define what we can call the **internal trunk**. They are involved in generally unconscious and involuntary vegetative functions (e.g. digestion) of the body.

The external and internal trunks are, of course, not truly separate because there are interactions between the somatic mesoderm and the lateral plate mesoderm. Further, the neural crest comes to influence both the external and internal trunks through migrations and differentiations that will be described later.

The Head

The initial development of the head is somewhat more complex than that of the trunk. That is why it is being described second. Within the head, the early development of the cranial end of the neural tube, eventually the brain, results in the head being the largest structure. This occurs very rapidly. The brain is even more complex, so that, again in accord with the idea of working from simplicity towards complexity, the brain and nervous system will be described last.

In the head as defined above, two sets of structures are apparent.

One of these results from the actions of a series of homeobox genes associated with the head ectoderm and **paraxial (somatic) mesoderm** cranial to the cranial end of the notochord. These tissues are associated with the general somatic functions of the head (and they parallel to some considerable degree somatic functions in the trunk).

The second are associated with special interrelationships between the ectoderm and the **visceral lateral plate mesoderm**. In the head this visceral lateral plate mesoderm is better called the **branchial** or **pharyngeal** mesoderm. It is not split into superficial and deep portions with an internal body cavity as in the trunk. Much of the lateral plate mesoderm in the head is intimately related both to the skin lying superficial to it, and the gut tube lying deep to it. Together they form pharyngeal or branchial structures.

These are as intimately associated with the exterior (for example, in relation to the development of a mouth and an auditory meatus) as they are the interior (for example, in a connection from the middle ear through a tube – the Eustachian tube – to the pharynx). The functions of these combined ectodermal, mesodermal and endodermal derivatives are much more conscious and voluntary than are the vegetative functions of the equivalent parts of the internal trunk.

Differences between Trunk and Head

That the trunk is not totally separate from the head is clear. Experimental deletion of the particular gene series producing the head eliminates only those head structures rostral to the hindmost part of the hind-brain. In other words, this part of the head has relationships with the true trunk. There are yet other divisions between the head and the trunk and within the head that we will examine later. In other words the delimitations of head and trunk are complex and fuzzy. Yet, though head and trunk seem to be different, both in structure and process, there are many materials and mechanisms in the development of each that are similar.

Structural Components

Thus, though in both the head and trunk, there are some similar building components, they have eventual different fates.

In the trunk these changes are most evident as a series of developments related to two main longitudinal structures: the longitudinally segmented paraxial (somatic) mesoderm and the unsegmented lateral plate (visceral) mesoderm.

Those components associated with the paraxial (somatic) mesoderm (including the more superficial parts of the lateral plate mesoderm – see later) will become the muscles, bones, joints and other tissues of the body wall. Their segmentation is evident in components called somites along the length of the trunk. The structures within them come to have somatic functions related to voluntary movements, conscious sensations and sensory information related to movements.

Those components developed from the unsegmented deeper portion of the lateral plate mesoderm (here termed visceral mesoderm) will become the muscles (smooth muscles, so-called involuntary muscles) and other supporting tissues of the digestive tube (and other structures developed from it, e.g. the respiratory tubes, the liver). Though they do not display segmentation like that of the somatic derivatives, they do become divided into a series of larger regions. They come to have visceral (or vegetative) functions related to the involuntary movements and unconscious information from viscera.

The junction between these two components is evident as the **serous body cavity** previously described. Only later will changes occur that separate this cavity into pleural, pericardial, peritoneal and other components.

In the head these tissue types are more complex. As in the trunk, they involve longitudinal series. But, in contrast to the situation in trunk, they are differently organized. The paraxial mesodermal components are segmented but the segmentation is into half units (somitomeres) rather than the full units (somites of the trunk). The equivalent of the lateral plate mesoderm (which in the trunk is not segmented) is, in the head, fully segmented, and this segmentation, too, is related to every pair of half units, the relevant somitomeres. This mesoderm is more generally called the branchial (pharyngeal) mesoderm and the segments are called branchial (pharyngeal) **arches** separated by branchial **clefts**. Additionally, there are, in the head, special sensory structures (**olfaction**, **sight**, **hearing** and **balance**) not present in the trunk at all.

The components associated with the head portion of the paraxial (somatic) mesoderm become the striated voluntary muscles and other tissues associated with (among other structures) one of the developing special sense organs (the external muscles of the eye) and the developing tongue (the tongue musculature). Those components associated with the branchial mesoderm will become the striated and voluntary muscles and other structures supporting the pharynx. Thus, in contrast to the trunk, both of these components include elements with voluntary movement and conscious sensation. The novel components involving the special sense organs are only evident in the head, the part of the animal that goes first into the environment.

Structural Mechanisms, Segmentation

More needs to be understood about segmentation. Another difference between the head and trunk in this early embryo relates to the development of segmentation in some of these structural materials. What at first sight seem to be a series of continuous structures from one end of the embryo to the other, eventually come to exhibit differences in segmentation between trunk and head.

Division into segmental compartments is a frequent mechanism in animal development for producing a variety of related structures. Thus families of what are called homeotic genes produce a repetitive series of generally similar compartments (segments). Such gene series are able to produce basically similar structures within each segment, but can vary the form of those structures according to the position of the segment along the row without requiring different genes, additional molecules or different mechanisms each time. The segments lie within what is called a morphogenetic (form-producing) gradient. There is, thus, a basic relationship among segmental structures within a segmental suite because they all arise through a single mechanism. Such basic relationships are what make adult anatomy easier to understand.

One early origin of segmentation in the embryo is, paradoxically, the unsegmented notochord (already described in the earliest stage of the pre-embryonic period). Developing between the ectoderm and endoderm in the midline, it forms the first supporting structure of the disk-like embryo. Though, itself, unsegmented, its early existence induces much of the later body segmentation. Once segmentation has occurred, the notochord gradually disappears. Some investigators believe that remnants of the notochord may be present as material between the vertebrae (in the nucleus pulposus of the intervertebral disks). This is, however, problematical because it is more likely that the original notochordal cells die, and are replaced by other mesodermal cells, a few of which may persist in the nucleus pulposus of the intervertebral disks. The real importance of the notochord is in its inductive effects in relation to segmentation.

The most obvious part of the segmental suite starts in the trunk (and is present also in the hind part of the head in a modified form). It is first evident in the paraxial (somatic) mesoderm on the dorso-lateral aspect of the embryo alongside the notochord. At first this segmentation produces many small elements that are called somitomeres. They become organized so that pairs almost immediately coalesce on each side to form somites. In other words, the somitomeres in the trunk are only fleetingly present. This coalescence reduces the number of somitomeres to give half the number of somites. These represent the beginning of the definitive segments, somites, of the adult trunk (Fig. 2.11(a)).

Later, a similar segmental suite also comes to extend part of the way into the paraxial mesoderm of the head. It is still material of somatic origin. However, in the

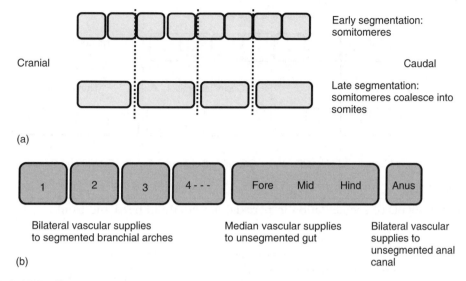

(a)

Bilateral vascular supplies
to segmented branchial arches

Median vascular supplies
to unsegmented gut

Bilateral vascular
supplies to
unsegmented anal
canal

(b)

FIGURE 2.11

Diagram of segmentation in the paraxial mesoderm (a) and in the lateral plate mesoderm (b).

head there are differences. In the hind part of the head, the somitomeres coalesce as
in the trunk to form somites (here called occipital somites). In the more cranial parts
of the head the somitomeres persist. They remain separate and so the segments that
can be identified are double the number in comparison to the eventual segments
that are found in the trunk.

The lateral plate mesoderm also shows different segmentation between trunk
and head. Thus the lateral plate (visceral) mesoderm that lies in the trunk on the
ventro-lateral aspect of the embryo below the notochord and along side the endodermal tube that will give rise to the endodermal derivatives (e.g. the gut) is not
segmented. In the head, in contrast, the lateral plate mesoderm is segmented. It
develops ventro-laterally in the branchial mesoderm supporting the endodermal
tube (that in the head will give rise to the pharynx). This segmentation is called
branchial (or pharyngeal) segmentation (Fig. 2.11(b)).

Finally there is also a later segmentation in the nervous system in the trunk so
that it is arranged as segmental peripheral nerves supplying the derivatives of each
somite. The transient somitomeres appear and disappear in the trunk before any
nerves are evident. The trunk nerves thus relate to the later somites.But in the head,
the brain is segmented very early through a series of neuromeres that coincide with
the head somitomeres (see below). There is, further, some evidence that the hind
end of the head is a somewhat intermediate structure, segmented like trunk somites,
but with nerves like those stemming from brain neuromeres (Fig. 2.12).

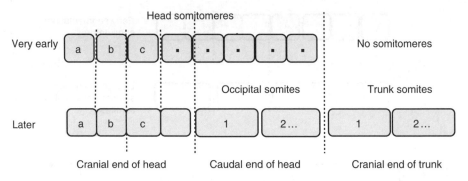

FIGURE 2.12

Further details of segmentation in paraxial mesoderm in head and trunk.

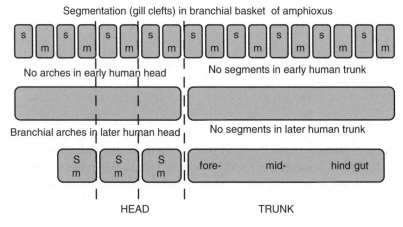

FIGURE 2.13

Further details of lateral plate mesoderm; also comparison with Amphioxus that has a branchial basket throughout the head and most of the trunk.

A similar differentiation between cranial-most head, more caudal head and trunk is also evident in the degrees of segmentation of the lateral plate mesoderm (Fig. 2.13).

Perhaps there really are three components: 'fore-head', hind-head and trunk. It is thus evident that the head is more complex than the trunk yet possibly earlier in its blueprint than the trunk

Structural Control, the Nervous System

The above leads on to the importance of the brain in understanding the body. In the trunk, the ectoderm gave rise to a midline cranio-caudal groove (neural groove) that later became a neural tube, the eventual nervous system. This tube shows, on each side, a division into (dorsal) alar and (ventral) basal elements (called plates) along the length of the neural tube. At the same time, the cell population in these plates, though not clearly separated into segments, do show evidences of segmental coalescences. These are in series with the somite segments described for the paraxial (somatic) mesoderm. Indeed, they may well be involved in the initiation of that segmentation.

These segmental coalescences are associated with the developing nerve fibers passing to and from the nervous system and the peripheral trunk tissues. Those passing inwards (conveying afferent information from the periphery to the center) stem from cell bodies that lie outside the nervous system in a series of structures, one for each segment, called dorsal root ganglia. The ganglia are, however, of neural origin, coming from the neural crest cells. Their central processes enter the dorsal alar plates in the spinal cord. Those fibers passing outwards (conveying instructions from the center to the periphery) stem from cell bodies lying in the ventral basal plates in the spinal cord. These separate sets of nerve fibers, called dorsal and ventral nerve roots, come to join within a single spinal nerve for each peripheral body somite.

The alar plates are primarily associated with somatic sensory inputs into the spinal cord and the sensory fibers enter the spinal cord via dorsal roots. The basal plates are primarily associated with somatic motor outputs from the spinal cord via ventral roots. These two roots (in each segment) become joined to form a mixed spinal nerve for that segment. Thus, though segmentation is not structurally as obvious in the spinal cord as it is in the periphery, it is every bit as present.

However, there are also sensory and motor fibers for the (lateral plate) visceral components. The viscero-sensory inputs lie with the somatic inputs in the dorsal roots, the viscero-motor outputs pass out with the somatic motor fibers in the ventral roots. These sets of fibers have a more variable distribution in the body tissues. Thus, though some of them do pass along the spinal nerves, many more are distributed as the mesh-work of nerve fibers in the connective tissues around blood vessels. Many more again are distributed to a series of nerve ganglia, also organized segmentally, but separate from the spinal nerves. These segmental neural elements are known as the right and left sympathetic chains.

Thus, in the trunk, to summarize: the sensory nerve fibers returning (afferent fibers) to the alar plates (dorsal) are conveying information from both structures that develop in the paraxial mesoderm (e.g. eventually the innervation of cutaneous, muscular, and connective tissue derivatives of the trunk) and information from structures that develop in the lateral plate mesoderm (e.g. the generally unconscious afferent innervation of both external and internal body wall elements).

Likewise, the motor nerve fibers (efferent fibers) leaving the basal plates (ventral) are conveying instructions to both paraxial mesodermal derivatives, such as the striated voluntary muscles, connective tissues and bones of the body wall, and as the generally unconscious efferent innervations of both external and internal body wall elements.

In the trunk these sets of fibers, separate at their entrances and exits from the spinal cord, come together (in humans) as the series of mixed (containing both efferent and afferent fibers) spinal nerves, one for each segment. The various types of fibers are distributed through the various branches of the spinal nerves to all the tissues.

In the head there is a similar alar (dorsal) and basal (ventral) division along the neural tube and, therefore, along the length of the eventual brain. The various sets of fibers exiting from the brain are arranged as in the spinal cord. The sensory information returns to the alar plates via dorsally located dendrites from nerve cells. The somatic motor instructions exit from the basal plate in ventrally located nerves axons.

There are, however, marked differences in the head. These relate to the way in which the various sets of fibers link up before entering (afferent) and after leaving (efferent) the brain. Thus the somatic motor fibers pass to paraxial mesodermal derivatives that in the head are the extrinsic muscles of the eye and tongue. They exit from particular somitomeres lying ventrally in the basal plate. They do not, however, join up with equivalent dorsal roots.

Further, also in contrast to the trunk, the motor fibers that are passing to branchial muscles derived from lateral plate mesoderm in the head exit the brain with the dorsal sensory fibers passing to the alar plates.

Likewise, in the head, as in the trunk, the nerve fibers returning to the alar plates (dorsal) are conveying sensory information from both the somatic structures (e.g. the cutaneous innervation of the head) and the visceral structures (e.g. the sensory innervation of the branchial (pharynx) region. These fibers lie with the same nerves already described as containing motor fibers stemming from the more dorsal in the basal plates and conveying instructions to the special muscles of the branchial arches.

Thus, in contrast to the trunk, the dorsal and ventral roots do not come together in what would otherwise be cranial homologs of the spinal nerves for each somite in the trunk. They remain generally separate, relating, rather, to the somitomeres in the head, so that there are many more cranial nerves than would otherwise be expected.

As a result, cranial nerves are therefore not equivalent to spinal nerves (see later). This is a situation that, paradoxically, given the apparent complexity of the head, may be actually more primitive than what currently exists in the human trunk (see Chapters 6 and 7). The cranial nerves mirror both the continued separation of somitomeres in most of the head and the segmentation of gut (pharyngeal) derivatives in the head. This is in contrast to the situation in the trunk where the spinal nerve arrangements mirror only the somitic segmentation (originally derived from early fusion of transient somitomeres into somites) of the trunk.

Thus, in both the head and trunk, but in different ways, the arrangements of the mesodermal structures are aligned with the arrangements of the nerves. The variations in these arrangements provide powerful explanatory hypotheses for understanding human anatomy.

From these initial developments, we can now see how the definitive trunk and head in the fetus arise.

2.3 Blueprints: Across the Chordates

The plans that we have just introduced in relation to development, occurring also in many other animals, mean that the human body is structured in parallel to the equivalent, but different body plans that can be discerned through the comparison of all those creatures possessing a notochord (known as chordates). The vertebrates are all chordates, but some chordates are not vertebrates; it is therefore this more inclusive term that we must employ. The word 'chordates' is already familiar to us. It names all those creatures for whom the notochord is the initial (sometimes even the only) supporting structure in the body. Though this book is about humans, a look outwards across the many other vertebrates, an even broader outward look encompassing the chordates, even, as a result, a look at some creatures that we would not easily recognize as being even chordates, is an indispensable scientific background to looking inwards to human structure.

2.3.1 Curious Beginnings

Thus the beginnings of body plans are evident in almost any living form that we care to examine. Thus, some creatures that are not even bilaterally symmetrical (jelly fish relatives – sea anemones) develop through similar genetic structures as humans. Both sea anemones and humans have Hox genes that are similar in structure and produce in both creatures their front and back ends (one cannot call them 'cranial' and 'caudal' ends because the sea anemone has neither a 'head' nor a 'tail').

Likewise, similar Hox genes, but many more of them, are evident in insects and humans. Here both creatures share many more Hox genes responsible for the production of heads at one end, tails at the other, and several components of the body in between.

These developmental mechanisms thus mirror those that are found in all animals with backbones (vertebrates). Nevertheless, even here vertebrate body plans are present in some very curious little creatures that we would not normally associate with humans. Particularly important are a number of animal groups that we can represent through some relatively small soft-bodied filter-feeding or parasitic marine animals known as Hemichordates (e.g. acorn worms), Urochordates (e.g. tunicates, sea squirts), Cephalochordates (e.g. lancelets) and Agnatha (without jaws – e.g. lampreys and hagfishes). The main bulk of the creatures in which we are interested, however, are the Gnathostomes (meaning with jaws openings – active animals with well-developed heads).

At this point it is important to remember that none of these creatures is a human ancestor. They are all extant species with their own long evolutionary histories. It is just that, in some of them, a degree of morphological stasis has characterized their evolution. Notwithstanding this, all these species have, as have humans, evolved since ancient times.

For example, though very far distant from humans, the Hemichordates do actually possess at least three homeobox (Hox) genes that are believed to be homologous to genes of the cranial cluster of Hox genes in humans. This reinforces the existence of phylogenetic remnants in human anatomy, especially the ancient nature of the human head.

It is possible to think of Hemichordates + Urochordates + Cephalochordates + Agnatha + Gnathostomes as belonging to an over-arching group whose body plans contain a notochord. The notochord is a very early element in human development (though it disappears with development). The Chordates also contain another extremely large subgroup: creatures with a vertebral column, called Vertebrates. Such a structure is also developed very early during human development, and though at first the vertebral column is a soft tissue structure, it eventually comes to be largely formed of cartilage and bone.

The Vertebrates, then, comprise (among others) cartilaginous fishes, bony fishes, amphibians, reptiles, birds and mammals. The last three are further separated as Amniotes. Amniotes are those vertebrates (reptiles, birds and mammals) producing embryos wrapped in a delicate transparent, sac-like membrane, the amnion. The differences in body plan among these creatures provide, perhaps surprisingly to the novitiate, important clues about what is fundamental and what superficial to the underlying human anatomy.

The Urochordate Plan

We can thus see that, long before there were humans, indeed long before there were vertebrates, there were chordates. All of the creatures related to humans, including of course ourselves, are chordates (that is they possess a stiffening element called the notochord, hence chordates) but some were chordates without a skull, without jaws and without vertebrae. In today's world these particular chordates are represented by a number of still existing species. Some of these are the tunicates. They hardly look as though they had anything much to do with humans. The adult tunicate (Fig. 2.14(a)) possesses one main element (that is also found in all chordates): a **pharynx** for dealing with the internal functions of food intake and waste product output. It is a sessile animal that seems to be nothing like any human, mammal, tetrapod or vertebrate.

The tunicate also has a larval stage, however, that possesses a mobile tail stiffened by a **notochord** (also possessed by all vertebrates at one stage or another). This makes it look slightly more like an early human embryo (Fig. 2.14(b)). This tail disappears in the adult.

The Pharynx

The pharynx, then, is the main structure of the adult. It is a bag that collects foods and gases from the surrounding water. It includes an iodine-concentrating organ called the **endostyle** (the human thyroid and some other tissues concentrate iodine). The pharynx is involved in digestion, leading into a simple very short digestive

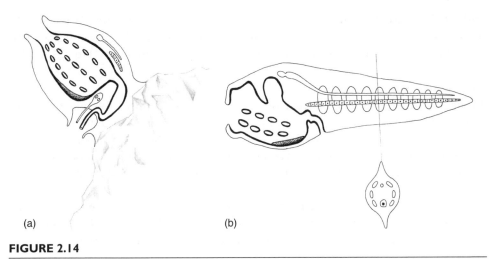

(a)   (b)

FIGURE 2.14

Diagrams of structures of an (a) adult tunicate and (b) a larval tunicate. The adult is, essentially, the larval 'head' minus its 'tail'.

system that consists of esophagus, stomach and intestine, with a terminal anus eliminating solid wastes that follow food digestion. The pharynx is perforated by slits so that water (that has entered via what we might call the 'mouth', the incurrent siphon) can exit by perforations (hence branchial slits) in a tunic (hence tunicates) surrounding the pharynx. This water exits to the exterior of the animal via an opening in the tunic called the excurrent siphon. During this passage, gases (such as oxygen and carbon dioxide) are exchanged. This therefore is a simple respiratory system. There are, of course, also gonads and a gonadal duct, a primitive reproductive system, lying close to the terminations of the very short intestine. All these structures are concerned with **visceral**, vegetative processes (such as food gathering, metabolic exchange and reproduction), and they are the principal structural elements of this creature.

The adult is thus a sessile form that sits on a substrate in the sea, filtering water and food with its pharynx. The animal can be thought of as an almost solely vegetative, or internal, or visceral, creature, yet capable of reproducing itself (Fig. 2.14(a)).

The Notochord
As indicated, the above is not descriptive of the whole animal. Tunicates have a larval phase that is motile within the sea. The larva consists of all the visceral structures enumerated above in the adult, together with a second set of structures. These comprise a tail-like component containing a supporting rod that looks somewhat like a short ventral notochord, together with associated muscle cells that can bend it, and a nervous system dorsal to it that drives the actions of the muscle cells.

A cross-section of the tail shows the dorsal nerve cord and the bundles of muscle cells around it. These structures are concerned primarily with swimming of the larva and can therefore be thought of as the locomotor, or external, or **somatic** parts of the animal. They largely disappear in those tunicates in which the free-swimming larvae metamorphose into the sessile, visceral, vegetative, adult tunicates (Fig. 2.14(b)).

The Neural Net

In the adult there is a small diffuse network of neural cells probably mostly related to the pulsations of the pharynx. In the free-swimming larval form there is a more complete neural net that is dorsal in position, develops from the overlying ectoderm, and is mostly involved in the movements of the tail. It scarcely deserves the term 'nervous system'. It is, however, more obvious than the tiny rudimentary nerve net of the adult.

The plan of the larva is fundamental in understanding this creature (Fig. 2.14(b)). This is what demonstrates its very distant relationship to the afore-mentioned cephalochordates, craniates, vertebrates, amniotes, and, therefore, humans.

Molecules

Finally, urochordates possess a number of Hox genes similar to those in humans. Other genes (Pax genes) pattern the neural elements, as they also do the equivalent neural tube in humans and other vertebrates. There are even present what are called BMP-like genes that in humans help set the boundary between the nervous and skin parts of the ectoderm. The presence of such common ancestor genes in such creatures is thus starting to emerge as an important theme in gene/anatomy interactions.

The Cephalochordate Plan

The cephalochordate plan (of which the lancelet is an example) is characterized by possessing the same two structural components, somatic and visceral, as the tunicate larva. In this case, however, both are present in the adult animal. The adult animal therefore looks a little like the larval form of the tunicate. It differs, however, in having a rather distinct head and trunk (Fig. 2.15(a–d)).

The Somatic System

This system is more extensive and more complex than that in the tunicate. It extends from the front of the animal to nearly the tip of the tail. It comprises first, a notochord running along this full extent. It includes, secondly, muscle associated with the notochord. The muscle is more complex than in tunicates, being organized as a series of segments along the body. The action of these segmented muscles together with the fact that the notochord itself contains myofibrils (thus it can change its degree of turgidity) produces lateral waves in the body wall. These waves, pressing against the water, mean the creature can swim in a more complex way than in the larval tunicate. Yet this structure is essentially equivalent to the somatic system described for the tunicate larva.

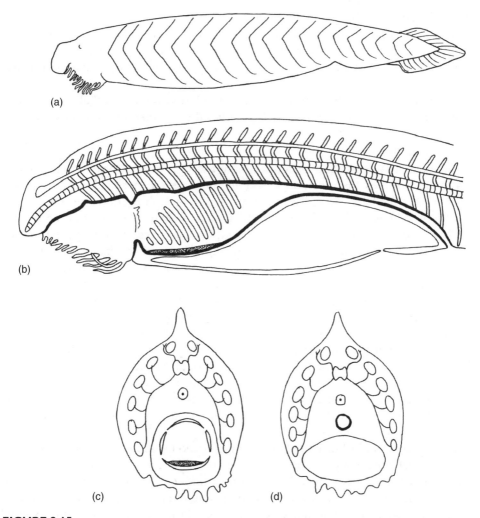

FIGURE 2.15

Diagrams of structures in a lamprey (a cephalochordate). (a) the complete external segmentation, (b) the cranially limited internal segmentation of the gut tube, and (c) and (d) cross-sections showing nerves at each of these levels.

The segmentation of the muscles (into what are called myomeres) is associated with a similar segmented pattern in the nervous system. The nervous system thus innervates the tissues via an equivalently segmental series of paired nerves, dorsal and ventral in their exit from the nerve cord on each side of the animal.

The dorsal nerves contain both sensory fibers from the whole animal and motor fibers to the non-segmented (visceral) muscle of the internal pharyngeal wall.

The ventral nerves innervate the segmented (somatic) muscles involved in swimming. These motor fibers are intra-segmental in location, passing directly between the nerve cord and the segmental muscles. They leave the nerve cord on its ventral aspect. In some species they are not really nerves, but rather myomeric muscle cells. They lie, however, in a ventral position similar to those of the true motor nerves found in vertebrates today; and they are equally involved in muscular contraction (Fig. 2.15 (a–d)).

This segmentation is also impressed upon the blood vessels so that, in each segment, a pair of segmental blood vessels supplies the myomeres and other structures.

The Visceral System
In cephalochordates like the lanceolate there is also a visceral system that is somewhat more complex than in tunicates. Some of these creatures, like the tunicates, are also filter feeders. Instead of a simple incurrent siphon, however, there is an actual mouth lying at the front end of the mobile animal. This mouth draws water and food into the next chamber, the pharynx, by ciliary action. In this creature, the pharynx is a large elongated structure encompassing perhaps half or more of the body length of the creature. As in tunicates, it contains slits in its lateral wall that communicate with the exterior (in this case directly), but, in contrast to the apparently random arrangement of the slits in tunicates, in the lamprey they are patterned as a segmented series of many slits from the front to the back of the organ. There is, again, an iodine-concentrating endostyle in the pharyngeal floor that is also associated with trapping food for digestion. Water, taken in through the mouth, exchanges respiratory gases with the blood, and escapes through the pharyngeal (or branchial) slits to the exterior. Food passes backwards into an esophagus, intestine and cecum, and food wastes exit through an anus. A small reproductive system is close to the termination of the gut. These various more caudal parts of the gut tubes, unlike the pharynx, are not segmented (Fig. 2.15(a–d)).

Though the Cephalochordate plan is characterized by its members possessing the same two elements, somatic and visceral, as the tunicates, they are, in these creatures, somewhat more complex. The somatic component is fully segmented from end to end of the creature. The visceral internal component is segmented at its extensive forward part (the pharynx, where the visceral parts are linked with external elements) but not along the small remainder of its length (the gut tube).

The Nervous System
These two sets of segmentation are also recognizable in the nervous system supplying these structures. These neural controls are also somatic and visceral, and are somewhat separated. Thus, simple though these creatures are, more of their components than in tunicates are like those in both developing and adult humans (Fig. 2.15 (a–d)).

Molecules
Finally, cephalochordates, or at least the lancelet that we are taking as our example, possess one Hox gene cluster with at least 12 genes arranged in the same order as the corresponding genes in mice and humans. In addition the lancelet has a number

of other genetic factors that are common to vertebrates including humans. It seems that the lancelet (though without jaws and a true head) contains homologs of those genes that do govern jaw and head development in vertebrates. Thus Pax genes pattern the neural tube, as they also do in humans and other vertebrates. Again, the presence of such common ancestor genes in these creatures and in humans (and of course other vertebrates), together with the way they seem to work, is emerging as an important theme in the basis of human anatomy. There is, however, much still to be learnt about all of this.

The Agnathan Plan

Those chordates still without a vertebral column but possessing a cranium without jaws (a = without, gnath = jaw) are yet another group. Examples are the lampreys and hagfishes. They exhibit all the body plan features of the urochordates and cephalochordates just described, together with a critical new feature: a cranium (of sorts) but still without jaws, and a trunk but still without a vertebral column. Of course, all vertebrates also possess a cranium but we are here looking at those specific forms that have a cranium but not jaws or vertebral columns.

The Trunk

All agnathans have a trunk possessing the various somatic structures showing segmentation. They all have a notochord and, though this largely disappears, true remnants persist in most of them into adult life. They all have a cranio-caudal series of segmented muscles associated with somatic functions and this segmentation is also evident in the neural arrangements. Likewise they all have the various visceral structures related to the internal tubes, the pharynx and the gut. As in the preceding forms, the cranial portion, the pharynx, is segmented, and the rest of the gut tube is not. But the segmentation of the pharynx is much less extensive, being confined to the head.

The Head

Thus, agnathans, compared with urochordates and cephalochordates, possess a further structure, a cranium, around the forward end of the nerve cord and gut tube. They do not have jaws associated with the entrance to the digestive tube (hence agnathans). From this point we can therefore use the term cranial for the forward part or direction in the animal and caudal for the tail part or backward direction.

The plan of the head includes the segmented pharynx and related structures lying ventral to the main nerve cord. These, together with new muscles that move the segments of the pharynx, assist in visceral, vegetative, feeding and respiratory functions of the pharynx.

Nervous System

The plan of the head is also related to an elaboration of the cranial end of the dorsally located nerve cord. As a result this forward end can now be termed a brain. In the prior creatures it was little more than the forward end of a poorly differentiated nerve net. This new elaboration occurs in part in relation to new sense organs that have evolved in animals that go forwards into their environment. These are primarily the olfactory, optic and stato-acoustic sense-organs (though there are also others).

Further and much less obviously, the plan of the head also includes the anterior end of the segmented somatic system characteristic of the trunk, though this does not extend all the way forward to the extreme cranial point of the head.

Finally, agnathans have two series of supporting structures in the head (but with no vertebral column). One of these lies superficially near the skin and surrounds the dorsal nervous system to form a cranium or skull. The other forms a deeper segmented series of structures supporting the wall and floor of the segments of the pharynx, forming a pharyngeal or branchial skeleton. This is also a plan that is associated with some of the elements of human anatomy.

Molecules

Finally, in agnathans, too, these various structures are developed through a series of genes (Hox genes, Pax genes, etc.) similar to those in humans.

The Vertebrate Plan

The term craniates, is an inclusive term. It includes the series of small animals that we have just described that have a cranium but no jaws; but it also includes the very large number of other creatures that have a cranium with jaws: fishes, amphibians, reptiles, birds and mammals (including, of course, humans) and they can be called gnathostomes (mouth-bearing jaws). However, unlike the previous species, these wide-ranging species also have vertebral columns. They are, additionally, therefore, also called vertebrates. Further, most of these species also share a series of other characters.

The Head

The most obvious of these characteristics is an elaboration of the cranial end of the pharyngeal slits. Here, the most cranial of the pharyngeal arches are segmented skeletal elements supporting and moving the opening for the entrance of food and water – the jaws (hence Gnathostomes – jawed opening or mouth). Other very obvious features are new adaptations in the sensory apparatus relating to the head. Thus the olfactory and optic organs are more developed. The double semicircular canal system of forms such as hagfishes is, in vertebrates, more elaborate with three semicircular canals.

The Trunk

There is a great elaboration of the trunk so that it contains along its axis a segmented vertebral column and related muscles for moving it (it is this feature that gives this group its name). Along the periphery of the trunk are two sets of paired structures (fins in fishes, limbs in tetrapods) and related muscles for movements of these structures.

The Brain

Finally, the simple nervous system is more elaborate so that a spinal cord and a three part brain are present.

These three developments, of trunk, head and brain might not be thought to be a particularly important series of new structures. However, jaws, two sets of appendages, better sense organs overall (and especially three semicircular canals in the ears) and the more complex nervous system gave these vertebrates special abilities. The jaws enabled them to sense, seize and eat a much wider range of foods. The two sets of mobile appendages allowed them greater ranges of movement within their environment. The sensory adaptations, especially those in the ears, gave them information about their position within their environment that further enabled them both to move and obtain food. The much more complex nervous system collated all of the above.

Molecules

Hox genes are particularly important in the production of the head. They coordinate the interactions between the skin and the neural crest, between the somatic and branchial segments, and within the developing brain (all see later). There is, moreover, a high degree of conservatism during development so that these components are relatively consistent among the vertebrates.

There are, likewise, a series of similarities among all these species in molecular determinants of the trunk. Specification within the trunk of vertebral regions, positioning along the trunk of limbs (in fish of fins) are all tracked in the patterns of Hox gene expressions.

2.3.2 The Overarching Comparative View

At this point we can then see that the major anatomical features characterizing humans are all present in the body plans of many other living creatures, and must have also been present in the evolutionary progenitors of them all. Figures 2.14–2.17 and 2.18 show some of the segmented aspects of the body plans of these creatures. Figures 2.19–2.21, and 2.22 show some of their internal visceral aspects.

The history of the study of the evolutionary relationships of all these animals is replete with competing and conflicting theories as to their evolutionary relationships. There are, also, many fossil species that add to the complexity. Yet further, even these short summaries make it clear that the new molecular studies of the living forms are adding to the understanding of these relationships. These comparisons are evidences of the enormous variety of the final products of evolution of developmental processes in many different forms. They also inform us, however,

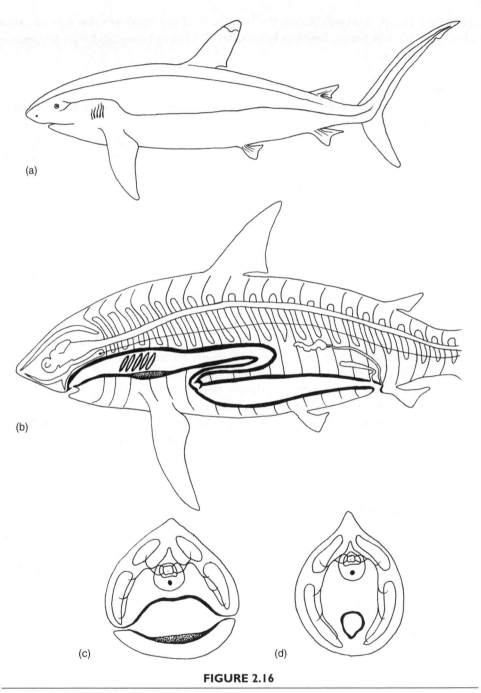

FIGURE 2.16

Diagrams of somatic structures in a shark.

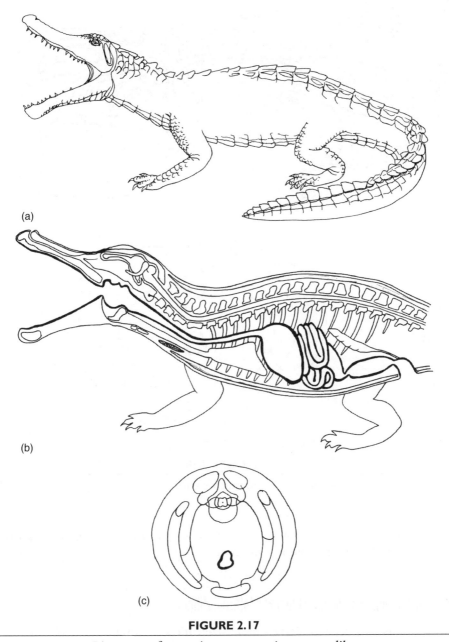

(a)

(b)

(c)

FIGURE 2.17

Diagrams of somatic structures in a crocodile.

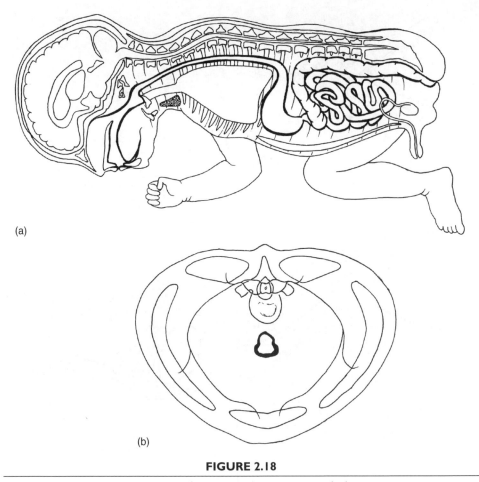

(a)

(b)

FIGURE 2.18

Diagrams of somatic structures in a baby.

about the common underlying mechanisms that gave rise to them. This is what is of value in understanding human anatomy.

2.4 Functions: External Lifestyles and Internal Milieux

The previous two sections have emphasized the developmental and comparative plans that can be discerned in the actual anatomical structures that exist in the different continuing stages and parallel forms. However, in providing those descriptions, it has been necessary to refer, many times, to function. Now we must pay more attention to the functions of the structures, especially the eventual functions as they exist in the human body. Again, let us start externally with functions relating to the external environment of the body.

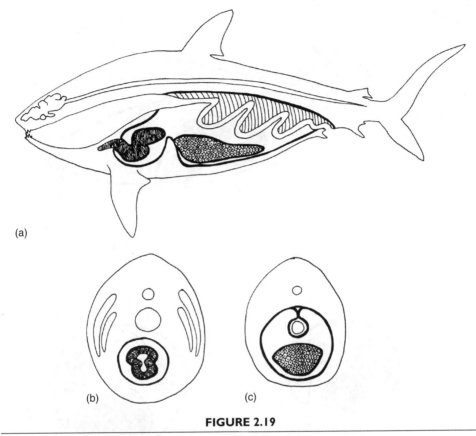

(a)

(b)　　　　(c)

FIGURE 2.19

Diagrams of visceral structures in a shark.

2.4.1　External Lifestyles: Somatic Functions

Just as we understand that there is a somatic structural component to the body, so we realise that there are somatic functions that it carries out. These relate to the lifestyles of the animals.

Movement in Water

Many of these creatures use their somatic functions primarily for movement in water. Of course, in tunicates it is only the larval form that is motile. This movement is carried out by the notochord and its attached muscle-like cells together with the nerve trunk that supplies them. The animal is very small and bendings of the notochord are sufficient, against the resistance of the water, to move the creature.

In other aquatic forms, the skeletal and muscular elements are segmented and thus, through differential contractions of segmented muscles from cranial to caudal,

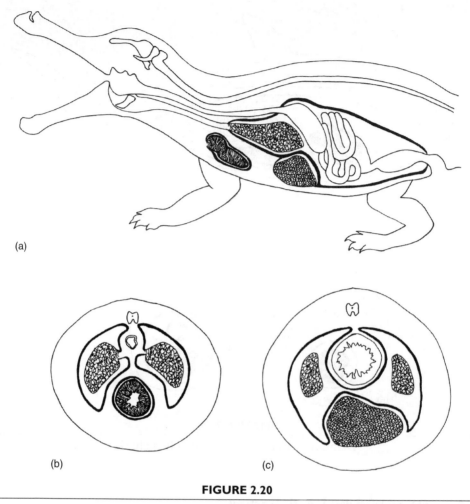

(a)

(b)

(c)

FIGURE 2.20

Diagrams of visceral structures in a crocodile.

greatly increase the lateral curvature of the segmented trunk. As a result, waves of contraction can pass caudally along the trunk, thus propelling the animal forward. This is the situation in lampreys.

In the cartilaginous and true fishes this lateral trunk bending is much greater and modifiable by appendages in the form of fins that help the animal control pitching, rolling and yawing. This greatly increases maneuverability within the aquatic environment.

Movement on Land

Fins (that controlled the three-dimensional movement in the water) are represented by limbs that have contacts with the two-dimensional surface on land. In many

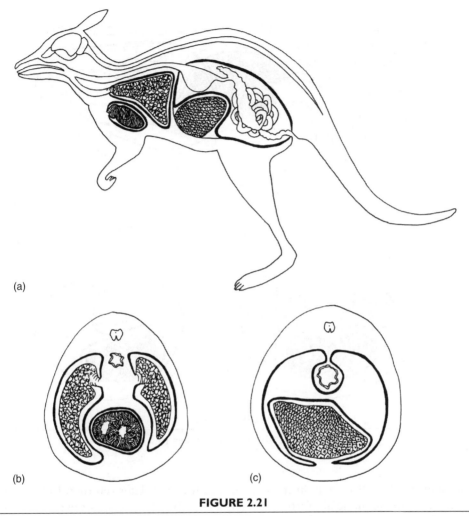

(a)

(b)   (c)

FIGURE 2.21

Diagrams of visceral structures in a kangaroo.

species, for example, many amphibians and some reptiles, lateral bending of the trunk (as in fishes) uses the contact points of limbs with land, rather than continuous trunk contacts with water, as a mechanism for making lateral waves become forward movement. Indeed, in many snakes lateral bending of the trunk alone suffices to allow the creatures to progress on land without limbs!

Locomotion on land further evolved so that, in many creatures, instead of lateral bending of the trunk being primary (though that still occurs), dorsoventral bending of the trunk becomes the primary truncal contribution to locomotion. This is combined with a shift from the primarily contact point function of the trunk or limbs driving lateral bending, to propulsive actions of the limbs as levers, driving the

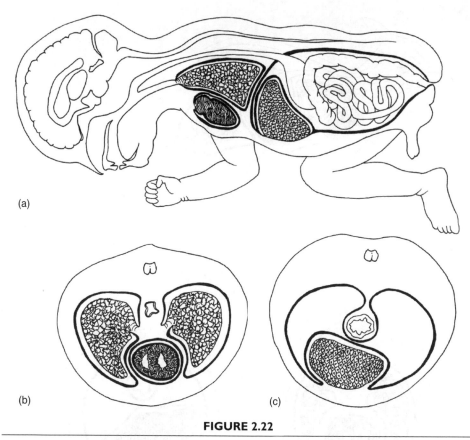

(a)

(b)   (c)

FIGURE 2.22

Diagrams of visceral structures in a baby.

body forwards. (Of course, in those land creatures that later returned to water, or had taken up life in the air, further changes occur; but they need not overly concern us in the understanding of human structure).

Finally, very recent developments in pre-humans have produced human bipedality, dependence upon hind-limbs alone for locomotion (although, again, there are many examples of other creatures that have independently evolved their own forms of bipedality).

The Place of Segmentation

All of these mechanisms originally depend upon segmented structures even though segmentation sometimes seems to be hidden. This is especially the case for the limbs. Human limbs do not look, from the outside, as though they are formed from many linear cranio-caudal segments. Such segmentation as they display (and this is not somatic segmentation) is proximo-distal along the lengths of the limbs: for example, arms, forearms, hands, thighs, legs, and feet. However, when we examine limb anatomies and even more limb functions, and especially limb development and neural control, then their cranio-caudal segmental basis becomes obvious.

Thus limbs receive nerves from particular segmental nerves (in what is called the somatic nervous system). The motor components of these give instructions from the nervous system (efferent) to motor elements (muscles, etc.) to produce the functions of locomotion and posture together with broader aspects, especially in humans, of non-verbal communication, tool making and handling, and so on. They control these movements, in part, upon the basis of information in sensory components (about the bodies external place in the environment and its internal operations) that are relayed back (afferent) to the nervous system as the somatic sensory system.

2.4.2 Internal Milieux: Visceral Functions

From Filter Feeding to Predation

All of these body plans have, like the human body, anatomical structures that deal mainly with the internal workings of the body, with what are often called visceral or vegetative functions. At this point we will take this to mean, primarily, the intake and handling of food, fluid and oxygen (though, as we shall see later, it also includes many other items). Related body components in all these developmental stages and comparative forms cope with these functions. In each there are anatomical parts, the gut including the pharynx (and their derivatives) that are involved in the intake and handling of food, and, at least in the cranial part of the gut, the pharynx and its derivatives, that are involved in the intake of oxygen. The pharynx specifically plays this double role.

As we have seen, in the various comparative body plans, the pharynx varies from being merely a multi-perforated sac passing food filtered from incoming water in tunicates, to a sac whose multiple perforations are segmentally arranged in the headless cephalochordates, to a similar sac that is incorporated within a head in the jaw-less fishes. In all other water-living craniates (fishes) the pharynx remains a perforated sac and is still involved in the intake and processing of food. The number of its segments are now much limited, and each segment has within it a series of skeletal elements. The first set of these skeletal elements become jaws for seizing and processing larger water-borne foods (though these are not yet the jaws that we see in mammals). The remaining skeletal elements relate to supporting and moving the gills that are gas exchange (or respiratory) structures. Even though the respiratory system changes on land, this structural arrangement remains the same in land vertebrates, though the pharynx becomes more complex, and additional skeletal elements come to lie external to those of the pharyngeal segments (pharyngeal arches proper).

The remainder of the feeding process involves food passing back into a non-segmented gut in the trunk with the detritus eventually expelled through an anus. There are various elaborations of that gut in the different body plans but these relate in the main to the complexities of the different diets of the different forms.

Many of the gut variations, therefore, are largely specially modulated elaborations adapted to particular foods and lifestyles. For example, creatures that are widely separated evolutionarily may, if they take large animal foodstuffs, nevertheless have a similar rather simple gut because handling animal foodstuffs does not require great complexity. Likewise, animals with lifestyles involving primarily plant foods all tend to display complexities of the gut related to the more complex nature of the processing of plant products. Such similarities, however, are usually parallel evolutionary developments associated with the specific environments and, therefore, lifestyles of the creatures at specific times.

Indeed, there are many examples of vertebrates that are extremely distantly related yet that have evolved remarkably similar structures in relation to special feeding behavior. For example, there are individual fishes, amphibians and reptiles (by definition, very distantly related) that have evolved 'gape and suck' feeding mechanisms. These involve mechanisms that permit the jaws to 'gape' at food materials, the pharynxes to 'suck' to draw them into the mouth, and the entire head structure to pass them caudally into the cranial end of the gut. Such similarities in these various creatures are completely parallel developments.

There are, alternatively, animals that are quite closely related (compared with the above example) because they are all primates (e.g. colobus monkeys – leaf-eaters, and tarsiers – insectivores) that nevertheless have totally different gut structures as adaptations to these different lifestyles. In other words, many of these differences in the modulation of body plans are not associated particularly with evolutionary relatedness or with human anatomy.

It is only in humans, and a very recent development it is, that differences in these structures can be related to a totally new form of feeding. This is one involving preparation of foods, sometimes a degree of external pre-digestion of foods, and chemical alteration of foods that all take place outside the body. Developments like these have had their effects in terms of reduced jaws and dentition, probably reduced digestive tubes, and reduced abilities to handle foods that are not pre-prepared in these ways.

From Water to Air

Parallel with these developments in relation to feeding, the pharynx also functions in relation to respiration, the intake of oxygen and the output of carbon dioxide. In the various aquatic forms the oxygen is contained within the surrounding water and is abstracted from it for use in the animal as it passes into the pharynx from the food entry portal. The carbon dioxide is eliminated as the water passes out of the pharynx through the lateral perforations. The complex blood supply to and from these pharyngeal arches and slits (gills) is heavily involved in transport of these gases to and from the body.

In initial conversion to land life, some creatures (e.g. many amphibians) carried out most of their gas intakes and outputs through the thin and moist body skin. Though they had gills for this purpose in their aquatic phase (the larva or tadpole), they were lost in the more terrestrial adult. Yet retention of an element of dependence upon water was maintained because these functions largely occurred through the moist skin. For us, this is a side issue.

With full conversion to land life, the pharyngeal mechanism still remains in the sense that the oxygen passes into the blood via the pharynx and thence to the animal's tissues. Carbon dioxide passes out of the tissues into the blood and then into the pharyngeal derivatives. These molecules, rather than being dissolved in water, are now gases in inspired and expired air.

With these new systems, the pharynx is still involved with food and air transmission. But these now become separated. The new air transmission passages (and, further, the gas exchange surfaces) have developed from the food transmission tube. Thus the cranial part of the pharynx proper leads both to the gut for food transmission and to the trachea to new gas exchange surfaces (the lungs). Air transmission tubes are thus now partially separated from the food transmission passages.

This almost certainly occurred more than once amongst different land vertebrates. It is always dangerous to assume that such developments are single ancestral happenings. Things confined to a single happening are probably very rare. (That is: if things can happen once, they can happen many times – the real exposition of *Occam's razor*. This implies that the simplest phenomenon is the most likely. However, the concept of things happening in parallel a number of times is far simpler than something, uniquely, happening only once). Thus, increasingly complex arrangements have evolved to separate air pathways from food pathways.

As a result of such developments, the cranial part of the entrance to the pharynx comes to have ventral oral (mostly food) and dorsal nasal (mostly air) pathways. These lead more caudally to ventral and dorsal pharyngeal structures. Associated with this, however, is a complex crossover. Thus the respiratory passages cross from dorsal (the nose) to become ventral (for example, larynx, trachea and bronchi). These tubes then lead to the respiratory exchange surfaces (lungs) that are ventral in position. The food transmitting passages cross from the mouth (ventral) to the dorsal food-transmitting passages (pharynx and esophagus leading to the stomach) dorsal in position. This is the situation in humans. Though it is an ancient and venerable evolutionary development, it is also involved in extremely new functions unique to humans, such as speech.

Excretion and Reproduction

A wide variety of anatomical structures have resulted from the beginnings of excretion and reproduction like those described in the simplest chordates. Again, these structures have probably evolved many times in different chordates providing several essays into egg laying, larvae, viviparity of various types, and so on. However, these functions in this great diversity in vertebrates relate mainly to the many different environments that the species have occupied: fresh water, brackish water, sea water, wetlands, dry lands, cold lands, warm lands, and so on.

2.4.3 Integration of External Lifestyles and Internal Milieux

One major controlling mechanism of both external lifestyle and internal milieu is the nervous system. It can be described as a central part, the brain and spinal cord, a coordination center, and a peripheral part, the nerves, the cables, as it were, that link coordination with function.

Let us examine the somewhat simpler peripheral nervous system first. As we have seen, the peripheral nerves are described as visceral and somatic in relation to the visceral and somatic functions described earlier. The system of peripheral nerves shows considerable complexity within itself, and also major differences between the head and the trunk. Let us first examine the peripheral nervous system in the trunk (because it is simpler).

The Peripheral Nervous System in the Trunk

In the trunk the nervous system is evident in the separate dorsal and ventral roots of the spinal nerves as they exit the spinal cord. These roots come together where these join, at each spinal segmental level, into a single pair of spinal nerves. Each spinal nerve at this point contains nerve fibers that are associated with external functions (**somatic**). Some of these fibers are involved with the external milieu, such as sensation (sensory information) and movement (motor instructions).

Other fibers of each spinal nerve are concerned largely with internal environment (**visceral**). Such fibers involve functions such as providing the nervous system with information about the pressure of blood (generally insensible information) and sending nervous system instructions for the contractions of smooth muscles and secretions of glands (generally involuntary motor and secreto-motor). Because of this complex of fibers, at this point these spinal nerves are called mixed nerves.

The somatic components are more complex than just **sensory** and **motor**. Thus in addition to obvious, conscious, sensory and motor functions, further fibers exist that are involved in somatic structures but in unconscious and involuntary modes. Thus, some sensory functions are unconscious in the sense that we are not aware of them (except when they go wrong). They nevertheless convey information to the nervous system about the states of somatic structures, the positions of joints, the activity of muscles, the orientations and movements of limbs, and so on. They are called proprioceptive (sensory) functions. Likewise, some motor functions are also unconscious in the sense that we do not consciously relate them to voluntary movement. Some, nevertheless, convey information from the nervous system to peripheral somatic structures (such as muscle spindles). These give information that is associated with the state and initiation of muscular activity.

These somatic components can all be thought of as the spinal portion of the somatic system of nerves, and they all evidence this through their segmental arrangement. They are largely conveyed to the periphery through the various branches of the spinal nerves. They remain fairly clearly segmentally arranged (each nerve contains components of only those particular spinal nerves that combined in its formation) and they supply muscles and skin in fairly clearly defined segmental blocks. However, the branches of these mixed nerves are best called **cutaneous** and **muscular** (rather than sensory and motor). This is because both contain motor and sensory fibers – as will be evident from the next few paragraphs.

There are, however, in the primary spinal nerves, a second set of fibers that supply those elements of the body that can be called **visceral**. These fibers, too, go to and from the nervous system. Those coming from peripheral structures provide information to the nervous system (and are called **afferents**). Others go to peripheral structures giving instructions to peripheral structures (and are called **efferents**).

The visceral afferents provide the body with information, such as from blood vessels (e.g. the state of dilatation of arteries) and from many types of glands (e.g. the secretory condition of sebaceous, sweat, and milk glands (the breasts – greatly modified sweat glands) in the skin). Of course, we are not generally conscious of these functions except in special situations, for instance, when we are aware that our skin is dry, or our breasts full.

The efferents provide the body with instructions to structures such as the smooth muscles in blood vessels (e.g. that blood vessel muscles should contract to produce vaso-constriction) and to glands (e.g. that sebaceous and sweat glands should produce their secretions). Again, of course, we are not generally conscious of these kinds of activities except in special situations, for instance, when we are aware that our skin has, that is, that we have, 'goose pimples' or 'ducky bumps' (through special activity of smooth muscles attached to hair follicles when we are cold or afraid), or are hot and sweaty (special activity of blood capillary networks and sweat glands when we are emotional or have increased metabolic rate).

This afferent/efferent system can be thought of as the spinal portion of the visceral nerve supply in the trunk. Unfortunately it is somewhat confused by terminology. Thus the efferent (motor) element of this is sometimes called the autonomic nervous system (and there are two parts, the sympathetic system and the parasympathetic system, but more of this later). It therefore seems to be differentiated from the afferent (sensory) components relating to those same structures. In fact both sets of components are merely the efferent and afferent loops of the same, visceral, arrangement. They are also segmentally arranged. There are, for instance a set of paired ganglia up and down the trunk – called the right and left sympathetic chains, and sets of unpaired ganglia in the head and pelvis – called parasympathetic ganglia – within this system.

The segmental arrangement in this part of the nervous system is somewhat disguised by the fact that these visceral components can meander up and down the body, sometimes supplying structures that are very far from the original segment. Even so, as we shall see later, there is somewhat of a logic to these wanderings that sometimes relates to the equivalent wandering of the structures themselves during development.

These visceral fibers, in contrast to those in the somatic system, are also distributed throughout the body in a more complex way. Some of them pass from the sympathetic chain back to the main nerves, and are thence distributed to the periphery through their cutaneous and muscular branches (as previously described). Others exit from the spinal nerves into nerve bundles, usually small and hard to find, that spring from the chain of sympathetic ganglia and pass thence also to the periphery. Perhaps the greatest majority are distributed in a yet different way, as fibers along the coats of blood vessels, supplying both those blood vessels themselves, and other structures that the vessels supply as they, in turn, pass into the periphery.

The Peripheral Nervous System in the Head

In the head the peripheral nervous system is far more complex. As we have seen, there are some nerves that are the equivalents (serial homologs) not of the trunk spinal nerves themselves, but of the roots of the spinal nerves in the trunk. But, whereas the spinal nerves are formed from the conjunction of the dorsal and ventral roots, in the head these equivalent nerve roots remain fairly separate as individual cranial nerves. Some of these nerves are, thus, the head equivalent of the somatic sensory and somatic motor nerve fibers of the dorsal and ventral roots of the trunk. Perhaps their separation in the head relates to the continued separation of the half-segments (somitomeres) in the head as compared with their fusion into full segments (somites) in the trunk.

Also in the head, however, are additional sets of fibers, forming additional cranial nerves that are not represented in the trunk. One set of these additional nerves is associated with the development of the new sensory organs, such as the nose, eye and ear, (and others!) that do not exist in the trunk. These sense organs are complex structures in which neural ectoderm and cutaneous ectoderm come into contact at special regions called sensory placodes. The nerves are primarily sensory and are called 'nerves of the special senses'. They are not technically nerves, being really brain extensions. The actual sensory 'nerves' are the very small olfactory fibers in the mucosa of the nose, the very short visual cells in the retina of the eye, and the very small hair cells in different parts of the internal ear. What appear to be the 'nerves' associated with these senses are in fact extensions of the brain and are better designated tracts, being essentially like other tracts that lie inside the brain.

A third set is associated with both the motor and sensory components of nerves that supply the special segmented structures of the pharyngeal arches that are not present in the trunk. As a result, they are called special visceral or special branchial (pharyngeal) fibers; special, because they are special to the pharynx and not present in the trunk; visceral because they are associated largely with internal functions; branchial (pharyngeal) because they are associated with the branchial (pharyngeal) apparatuses. Yet the information they are concerned with, the sensations they subserve and the movements that they produce, are usually very conscious and very voluntary, sensations of which we are aware (e.g. taste), or movements that we intend to make (e.g. swallowing).

Though they have a certain equivalence with the visceral nerves in the trunk just described, they carry out far more complex functions in relation to the complicated head end of the embryo. The conjunction of external and internal functions that they cover can in no way be described as unconscious and involuntary. Although some fibers within them are indeed unconscious and involuntary, the great majority are associated with highly sophisticated conscious and voluntary activities of the head end of the gut tube, that component of the head that is a conjunction of ectodermal and endodermal elements.

Finally, there are sets of fibers that arise in the head, but that pass caudally into the trunk, as special branches of certain cranial nerves, to supply fibers to many trunk structures. They are called the cranial parasympathetic nerves and are another component of the autonomic nervous system. Again, as before, this term is unfortunate, seeming to include only the efferent (instructions) fibers. But within them is an equivalent set of afferent (information) fibers. This system is generally unconscious and involuntary. On occasion however, their functions obtrude into the conscious/voluntary universe, as when we realize the fluttering of the heart in disease or emotion, or feelings of extreme cold in conditions of fear. These afferent and efferent components are very important in the functions of many body structures (such as the thoracic heart and lungs, the abdominal gut and organs). Interestingly, most of the pelvic organs are also supplied by a parasympathetic system. This has, however, little to do with the head and is a special structure of the pelvic part of the abdomen (see later).

The peculiarities of the distribution of many of these autonomic fibers are, in part, because many of the trunk structures that they supply have, originally, developmental and evolutionary origins in relation to the head. For example, the thoracic heart and diaphragm originate in the embryo far cranially, even in advance of the developing brain. They gradually migrate caudally through a ventro-caudal folding of the head, into a region that will be the eventual thorax. The lungs originate from the pharynx in the head but also later descend into the thorax. Even the thyroid starts off as a development of the endostyle in the floor of the pharynx and only later comes to have components that have descended into the neck (thyroid) and even into the thorax (mediastinal thyroid) and even caudally as far as the stomach (which may also contain iodine-concentrating cells) as variations in development in some individuals.

2.5 Integration and Control, Body and Brain

In all of what has passed before, we could easily be forgiven for thinking that the study of brains was separate from the study of bodies. Indeed, in textbooks they are generally treated separately. Yet whatever the relations are between brains and bodies, it is evident that both are present from the beginning, the beginning in both evolutionary and developmental terms.

It is equally obvious, but much less clearly understood, that, through development and evolution, there are many reciprocal interactions between body and brain, brain and body. Elements of the brain are involved in the arrangements of body parts. Elements of the body are associated with the arrangements of brain parts. This duality is why the nervous system (normally the province of the separate discipline, separate textbooks, of neurobiology) also has a major place in the understanding of the anatomy of the human body.

2.5.1 Inside the Nervous System: Spinal Cord in Trunk, Brain in Head

Although the peripheral nervous system has already been described in two portions: for trunk and head, respectively, and though the central nervous system can also be so described, in fact the central nervous system is considerably more complex than just a bipartite structure. In its organization, moreover, as with the differentiation between trunk and head, the separation of parts is not clear. In the head, it is as though there were a degree of blending from simpler mechanisms in the trunk to more complex ones in the head as one passes from caudal to cranial. Thus there is further complexity as we look cranially from the caudal (occipital) end of the head towards the cranial (rostral) end.

All this also applies to the central nervous system with parallel differences between spinal cord and brain, and greater complexity within the brain. This complexity, is, finally, present in spades within the cranial-most part, the fore-brain. These increasing complexities relate to the primary controlling and integrating functions of the central nervous system.

Thus a portion of the central nervous system involves components that lie on each side, supplying the body segments at that level and, therefore, integrating structures and functions within each segment. Other internal portions are organized as transverse elements that link, and therefore integrate, the segments of the two sides. And there are yet other components, longitudinal connections, which link, up and down the body, the various segmental levels. Finally there are, in the head, additional new structures apparently not related anatomically to any segmentation.

2.5.2 From Body Segments to Peripheral Nervous System Segments in the Trunk

Variations in body segments in different creatures are related to variations in their nerve supply.

Very ancient patterns

Creatures such as today's lancelet have a series of paired 'dorsal' spinal nerves. These contain three kinds of fibers: sensory from the skin, muscle and connective tissue (hence somatic sensory), sensory from internal organs (hence visceral sensory), and motor to visceral effector organs (hence visceral motor). These are intersegmental, passing to their structures along the connective tissue sheets (myosepta) that lie between the segmentally arranged muscles. All nerve bodies of these 'dorsal' spinal nerves are located within the spinal cord so that there are no 'spinal ganglia' as there are in humans.

There are also, in the lancelet, effector nervous structures that are segmental in position and these were formerly called ventral spinal nerves. They are, in fact, not nerves per se, but segmentally arranged specialized muscle fibers that control animal movements (and so could be termed 'somatic'. These muscle fibers run back from the muscles in the periphery to the surface of the spinal cord centrally where their motor endplates are located. In the lancelet this is a very specialized situation. Nevertheless, it gives us a first glimpse of the separation of affector reception from effector control.

Going with the alternate arrangements of these 'nerves' are alternating dorsal and ventral segments of the body. As a result, to cope on each side with one pair of nervous elements (dorsal and ventral) are two 'segments'. It is not clear what are the time relationships of the 'nerves' and 'segments' – whether the one comes before the other or vice versa.

In lampreys the dorsal spinal nerves are like those in the lancelet except that some of the dorsally located sensory neurons here have cell bodies outside the spinal cord. There are true ventral spinal nerves in lampreys (instead of ventral specialized muscle fibers) that supply muscles in the body (hence truly somatic motor) and these fibers are totally external to the spinal cord. As in the lancelet the dorsal and ventral elements do not join, but run independently as separate dorsal and ventral nerves to their destinations. And again, as in the lancelet, there are alternating dorsal and ventral segments in both the body and the spinal cord.

Modifications of Those Ancient Patterns

Further variation exists in hagfishes, fishes and amphibians. The dorsal and ventral nerves (at each level and on each side) join outside the spinal cord to form combined spinal nerves. There is, thus, a single pair of spinal nerves in series along the trunk. The nerves that join to produce this spinal nerve are now called the dorsal and ventral roots (of the spinal nerve). The join is outside the vertebral column that has developed around the spinal cord. There is a further difference: viscero-motor components exit in both dorsal and ventral roots.

In relation to this junction of the dorsal and ventral nerves, there is also a junction of the body segments, so that the number in hagfishes, fishes and amphibians is half the number in lancelets and lampreys.

The Body Plan in Amniotes Including Humans

Further modifications of the plan are evident in many amniotes including humans. The 'half-segments' described in hagfishes, fishes and amphibians, are only very transiently present in amniotes. In amniotes, almost immediately, before the development of spinal nerves, the half segments (called somitomeres) give way to half the number of full segments (called somites). It is at this stage that the spinal nerves of the trunk appear so that these are never anatomically related to the somitomeres – they are immediately related to somites. (It is, however, entirely possible that molecular factors related to the half-segments are involved in the production of the spinal nerves.)

There are further differences. The dorsal and ventral roots join together as a single segmental nerve, as just described for hagfishes, fishes and amphibians, but in most amniotes, the viscero-motor components are confined to the ventral roots. There has thus been a shift of viscero-motor fibers from dorsal in lampreys, to both dorsal and ventral in fishes and amphibians, and to ventral alone in amniotes (which of course, include humans). This leaves the dorsal roots in amniotes including humans containing only sensory neurons (both somatic and visceral).

Thus, in the trunk, in humans, the anatomical segmentation of the body structures (somites) comes first, the anatomical segmented arrangement of the spinal nerves second. It is entirely possible that this is not true at the molecular level; the molecules may very well predate the production of the spinal nerves. It is even possible that, aeons ago, the same was also true in very ancient ancestors.

2.5.3 From Body Segments to Peripheral Nervous System Segments in the Head

Let us now look at the implications of this for the head and its cranial nerves. The spinal situation is not true for the head.

Some differences in cranial nerves stem from the nature of segmentation of the head and brain. In the trunk in humans, as we have just seen, somites are already present in the periphery in the embryo when the nerves are not. Thus the segmental arrangement of the spinal nerves is regular and may well have been secondary to the very early segmentation that produced the somites.

In contrast, the head is more complicated. In part this is because those 'half segments', (somitomeres) that were only transiently present in the trunk, are fully present in the head and, except in the occipital region, do not go on to fuse into full segments (somites). The occipital region does seem eventually to have segmentation somewhat like that in the trunk (some head somites with dermatomes, myotomes and sclerotomes). However, this does not occur within the brain more cranially; the pattern of 'half' segments is maintained.

Secondly, an independent series of segments not found in the trunk of amniotes (even though present in the trunk of the lancelet and the lamprey) develop in the head. These are the derivatives of the branchial (pharyngeal) structures.

Thirdly, some head components are formed that may have been present before even segmentation arose in evolution. These are special sensory components such as the olfactory and optic sense organs and nerves.

All these affect the cranial nerves and are perhaps related to the fact that the 'half segments' do not join to form 'full segments'. Thus the dorsal and ventral nerves at each level do not join to form a single nerve as do the dorsal and ventral roots in the spinal nerves passing to trunk somites. They remain largely separate. In other words, this situation in the head seems more primitive than that in the trunk – more reflective of what exists in the adults of such forms as lancelets and lampreys.

As a result, certain of the cranial nerves are in series with the dorsal roots of the spinal nerves, though they more resemble the spinal dorsal roots in lampreys than the dorsal roots of the human spinal cord. That is, as in lampreys, they also contain viscero-motor elements. These pass to the muscles of the branchial arches, which are not present in the trunk.

A second group of cranial nerves are in series with the ventral roots of the spinal cord in humans, but are also more like those spinal ventral nerves in lampreys. These nerves innervate derivatives of ventral head somitomeres (and in the hindermost part, occipital region, of derivatives of what truly resemble somites, occipital somites). In other words, they supply special muscles, the muscles of the eye and (in the occipital region) tongue that are somatic in origin.

Finally, a number of cranial 'nerves' exist in humans that have no counterpart in the spinal series in humans. These innervate structures that relate to the several sense organs that are peculiar to the head, the part of the organism that goes first into the environment. It is common to call these 'nerves' 'special sensory' or 'special somatic sensory' in recognition of their distinctive nature in the head. They are also sometimes called 'suprasegmental' in the notion that they lie above the segmental structure of the head. However, these 'nerves' are actually tracts of the brain that lie outside the brain; they too are segmentally derived; they derive in relation to special neurectoderm–ectodermal junctions called placodes. The nature of this evidence is, again, at the molecular rather than anatomical level.

Of course, the cranial nerves are numbered in the spatial sequence in which they exit from the brain in humans; but the above exposition makes it clear that these numbers are quite arbitrary (indeed, actually confusing). The numbers were given at times when some of the nerves had not been found (by human anatomists). They were also given when some of the nerves were not recognized as really tracts of the brain lying, like nerves, outside the brain. They were also muddled because they were numbered in a sequence of location in humans that did not (could not in those days) take account of bends and folds in the human brain.

2.5.4 From Peripheral Nervous System Segments to Central Nervous System Segmentation

The question that then arises is: how is segmentation in the central nervous system related to segmentation in the peripheral nervous system? The answer, not totally understood, is again different in trunk and head.

The Spinal Cord

In that component of the central nervous system in the trunk (the spinal cord) there is no early developmental evidence of those transient initial 'half-segments' that appear in the trunk. And even when, in the trunk, these fuse giving rise to the primary trunk segmentation, the somites, equivalent anatomical components are not yet present in the spinal cord. It is only later that the spinal nerves formed from the fusion of dorsal and ventral roots start to migrate out into the trunk and divide into their various branches. Of course, it is not impossible that these elements are indeed present at very early stages in the nervous system, but if so, it is at a molecular level not anatomical, and has not so far been identified.

The Brain

In the brain, however, the reverse is the case. Not only are the 'half segments' present in the periphery of the head, and are related to equivalent cranial nerve structures migrating from the brain as just described, but also the brain itself evidences internal 'half segments'. The term we have already identified to recognize this outside the brain is the word 'somitomeres', a modification of the word 'somites' for full trunk segments. Inside the brain, the overall term is 'neuromeres', but as these are generally recognized in relation to the major brain parts, they are given different names in the different parts. They are called: 'rhombomeres' in the hindbrain (numbered caudally from the cranial limit of the hindbrain, r1 to r8); 'mesomeres' in the midbrain numbered caudally from the cranial limit of the midbrain (M1, M2 and the Isthmus, this latter term handling a 'boundary dispute' between the midbrain and the hindbrain); and 'prosomeres' in the forebrain (numbered, just to cause further confusion, in the reverse direction, from the caudal limit of the forebrain to its cranial point, P1 to P6).

Associated with these various 'neuromeres' are a series of molecular factors that developmental biologists have shown are fundamental to the construction of both trunk and head.

In the spinal cord and brain the ventral structures are influenced by molecular factors, such as 'sonic hedgehog' secreted by midline structures (in the trunk of course, by the notochord, and in the head by the notochord as far forward as it extends, but also even further forwards beyond the cranial limit of the notochord).

Likewise in the spinal cord and brain, the dorsal structures are influenced by bilateral factors such as Wnt. These influence the production of the dorsal (in the spinal cord) sensory structures, but dorsolateral (in the brain) sensory structures, together with the elaboration of the special large dorsal structures, the cerebellum and the cerebrum that are not found in the spinal cord.

These complex segmented structures in the brain parallel those in the head outside the brain and, indeed, may eventually be found to be precursor developments. Thus, despite the novel complexity of the head and brain in humans, they reflect both head and brain structures that are much older, phylogenetically speaking, than those of the trunk.

Though somewhat complex, understanding these fundamental patterns aids enormously the understanding of human anatomy. We will go further into these complexities in a later section on the central nervous system.

2.6 Evolution: Forwards from Deep Time

Contrary to what we might expect, we do not yet have anything approaching a single 'consensus' evolutionary tree that includes the human species. I am sure we never will because we will never know all of the animals that have ever been fossilized. Even those fossils that we do have, though seemingly many, are only very few compared to the total numbers of species that have ever lived. In fact, we do not even have an inventory of all the animals alive today although we do know most of the larger vertebrates. A new mammal is occasionally, even now, recognized from time to time.

Those creatures we know as fossils are largely fragmentary and their anatomies are usually only represented by bits of the hardest materials, shells, bones, teeth and so on. There are, of course, some remarkable exceptions: fossils showing feathers, skin, scales and even fur on the outside, stomach contents, parasites and fetuses on the inside, and even, occasionally, information about ancient molecules (e.g. DNA) and atoms (e.g. enclosed metals). It is thus becoming increasingly possible to find some genetic information about some fossils.

2.6.1 Human Anatomy and Evolution

Perhaps surprisingly, the anatomies of those fossils alleged to be rather like humans do not provide a consensus picture though one would think, from most popular versions of human evolution, that it was all settled. Almost as many evolutionary trees leading to humans have been proposed as there are fossils to be sorted onto them. Further, we do not seem to have found (and in my opinion also surprisingly) anything much that is **not** purported to lie in the human part of the evolutionary tree. For instance, there is only a single pre-chimpanzee fossil (and that only a few teeth) in comparison to the very large number of alleged pre-human fossils in the same time period! And we have found no fossils at all, that anyone is willing to recognize as neither pre-human nor pre-chimpanzee! Always, it seems, we want to place every new fossil in the pre-human part of the relationships, often, as the earliest and most critical, find.

However, that evolution has occurred is a fact. How it has occurred, and what is the impact for human anatomy, is the puzzle. One of the problems for understanding evolution and its mechanisms and processes stems from the fact that the most recent developments in the field are not well known and are frequently confused with prior ideas.

For example, very early ideas about evolution used the metaphor of a ladder, implying that evolution is a procession up a ladder of species, with, of course, mammals, then primates, then humans at the top (Fig. 2.23(a)).

A later idea used the metaphor of a tree with earlier forms on trunks and in the broad canopy, with present day forms at the ends of branches, with, of course, humans highest at the crown. This is a somewhat better representation, illustrating

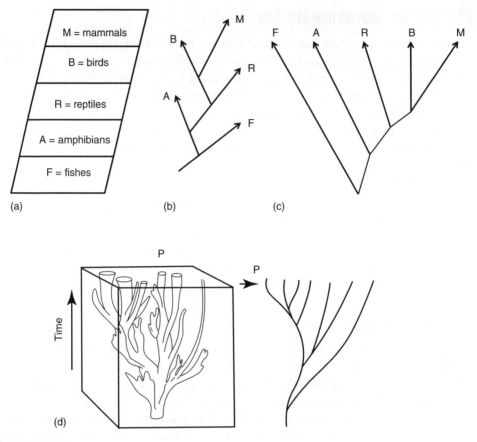

FIGURE 2.23

Some metaphors for evolution. (a) ladder of life, (b) tree of life with humans at the apex, (c) two-dimensional bush of life with all present day forms at the same level, and (d) three-dimensional tree of life that can be derived from the two-dimensional bush of life.

the order in which different groups appeared, but is still misleading. No present day species is 'higher' than any other (Fig. 2.23(b)).

Much current usage employs the metaphor of a bush, with a complex of small twigs, and where there is no special place for any given present-day species (Fig. 2.23(c)).

Evolution is perhaps even better represented as a three-dimensional volumetric structure. The positions of given species are represented at each time slice, and the separation of a species into two can be shown over time. The cross-sectional area of the structure allows us to include the concept of variations and differences across a geographic region at any given time level, the volume to include variations and differences continuing through regions and through time (Fig. 2.23(d).

These increasingly sophisticated metaphors are attempts to demonstrate how the process effects variations in individuals within a species, and differences in species within higher groups. Yet, of course, the relatednesses among individuals and the relatedness between species, though intertwined in evolution are, themselves, quite different phenomena.

Relatedness among Individuals

The degrees of relatedness between individuals might be expected to be reflected by some portion of the biological information about the individuals. What is meant by individual relatedness? It can obviously be explained thus. Brothers and sisters are more closely related than are cousins. Brothers and sisters share parents in the immediately prior generation. Cousins share parents in an earlier generation. When we do this we are looking at genealogies. Although we only rarely have full genealogies, we do have some good examples, especially in following up rare medical conditions, such as hemophilia in Queen Victoria's relatives. In such a case we can actually know who is related to whom.

This determination of genealogies is currently undergoing a marked improvement as a result of modern molecular comparison technologies. They give rise to an ability to track true relationships of individuals back into the past. Even so, however, when we do so we get some surprises. For example (Fig. 2.24(a)), using a simple genealogical diagram, which is not like a tree or bush because branches join as well as separate, we can track back the mothers of all the people in the present generation (the top level of the diagram) until we find the ancestral 'mother' further back at some lower level in the diagram (notice, time is in the reverse direction, from the present above to the past downwards in such a diagram). This can specifically be done in practise by tracing mitochondrial DNA (passed on to both sons and daughters from mothers). An equivalent analysis can be done for Y-chromosomes (different because Y-chromosomes are passed only to only sons).

When, however, we look at the genealogy implied in Fig. 2.24(a), but work out the relationships as we would have done if we did it a couple of generations earlier, we get a completely different answer (Fig. 2.24(b)). The ancestral mother of everyone in that generation is not the same 'mother' as is found in tracing back from the top (present) generation. She is actually much further back. In fact, as can be seen from the diagram, because it is limited to only a few generations, we never actually locate the 'single mother'; she is earlier than the start of our small example. There are, in fact, two such mothers in our 'earliest' generation.

Of course, the reason for this is obvious – doing the analysis from an earlier generation includes individuals in that generation that did not send offspring into the next generation, and so these individuals do not contribute to finding the ancestral mother for the later generation. Multiply this concept over a much larger number of individuals, and many more generations, and one can see how it is possible to be enormously misled. This is especially so when large numbers of individuals

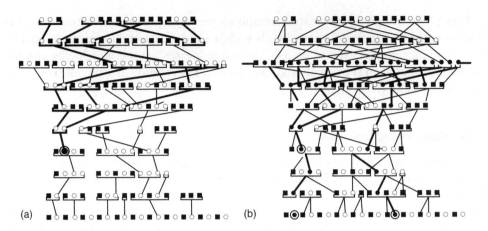

(a)   (b)

FIGURE 2.24

Ten levels of a genealogy from individuals at the present day (squares = males, circle = females) with the present day at the top of the diagram. The ancestral mother (circled individual) is seven generations back (a) when we calculate from the present day. However, if we start from three generations back from the present day (b), then two ancestral 'mothers' (circled) are defined. Both of these are in the initial generation of the diagram, and this means that the single true ancestral mother is even further back in time. Note the differences in the two ancestral mothers depending on which generation one works back from.

(as during wars, epidemics or famines) are lost in each generation. This simple discussion makes us realize just how time-bound such analyses are.

Relatedness between Species

When we come to apply such ideas to species of individuals, rather than to the individuals themselves, the matter is different. We are no longer looking at genealogies among individuals, but at lineages between species.

Individual genealogies (alternating patterns of births, matings and births) diagram a process of union then separation, union then separation, following in each generation over time. Species lineages show the continued separations of species over much longer time spans. Usually we envision species as continuing unchanged, splitting into two, or going extinct. (We can, however, recognize the situation where two species (if still capable of interbreeding) or subspecies (if so designated) can undergo union back into one. This is not, however, similar in any way to the situation for individuals.

Again, let us look at a simple example. One might think that, if two present day species, A and B, share a more recent common ancestral species than present day species C, then species A and B should be more similar to each other than they

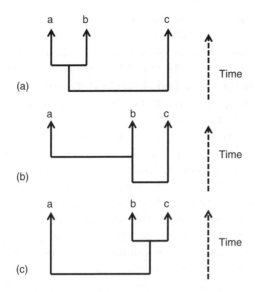

FIGURE 2.25

These diagrams show the relationships between three species A, B and C, under different conditions of 'structural difference' (horizontal axis) and 'time change' (vertical axis). The first frame (a) indicates a degree of structural difference giving the same relationship as the degree of time change. The second (b) and third (c) frames indicate how differences between the degrees of structural difference and the time change can give two other sets of relationships.

are to C. This may in fact be true (Fig. 2.25(a)) so that time change and structural difference can be associated with one another.

Let us suppose, however, that species A has changed a great deal since the time of separation from its sibling species, B. In contrast species B may be little different from C because neither of them changed much in the overall intervening period of time from the initial common ancestral species. This provides a diagram of relationships as shown in Fig. 2.25(b) where time and difference are disassociated.

If, however, we do not know that species A has evolved at a much faster rate since its separation, then examining these differences, assuming association, will give us the wrong result (Fig. 2.25(c)).

From Individuals to Species

Obviously there is a point at which massed genealogies of individuals in a population (a single species lineage) can give rise to two lineages of two new species. How does this relate to genealogies of molecules in a population (of a single species) giving rise to new molecules in two new species? Are the individual lineages and molecular lineages the same? Do they give the same picture of relationships? The answer is that they are very unlikely to do so.

Thus Fig. 2.26(a) shows a situation where a population of individuals (shown as the cross-section of the lower wide trunk) gives rise (upwards) to two new populations of individuals (two new wide branches). One would normally think of the level of the split as being the start of the two new species.

However, if we trace backwards from the present day (the top of the diagram) in order to find, shall we say, the ancestral mothers for present day individuals in each new branch, we might easily find that they were (a) separate for each, and (b) that the original mother of them both did not occur at the split, but much further back in time. If interbreeding after the split is not occurring, the original population trunk represents a species that has become two species. Determining the ancestral mothers, does not necessarily reflect the species split.

A modification of this figure (Fig. 2.26(b)) shows that splitting of lineages of genes does not (necessarily) coincide with splitting of species. Moreover, different genes trees split at different times. Just as the fact that the starting time of genealogies may influence the results for populations, so too may the choice of particular genes.

In other words, neither genealogies of individuals nor genes are related directly to the splitting of species.

These ideas are especially relevant to the species (humans) in which this book is particularly interested. And the example is rendered more complex by the fact that one can carry out such analyses using differences based upon human anatomies or human molecules, and upon times based upon both anatomical and molecular calibrations. The anatomical time calibration can be produced by assumptions from fossils. The molecular time calibrations can be assisted by assumptions about rates (though, more and more, nowadays, information about real fossil molecules is being discovered).

2.6.2 Humans and Apes

Let us apply some of this thinking to humans. Thus, an older view of human evolution based upon an assessment of the anatomy of humans (species A) saw them as more different from both chimpanzees (species B) and gorillas (species C) than these apes were from each other. This view, thus, placed the African apes in an ape family, and humans in a human family. With this went the assumption that, though humans and these apes were accepted as having a common ancestor, that common ancestor must have been a very long time ago. Darwin didn't even try to put a time on it. Workers after Darwin's time, and on into the first half of the 20th century, did use such data (including a few fossils not known to Darwin) to provide times of common ancestry. These times, however, were often very long – 20 to 30 million years.

However, by the middle of the last century, comparisons of fossils not known to the earlier workers (e.g. australopithecines) and new biochemical ideas (e.g. immunological relationships) started to produce a markedly different view. This reckoned

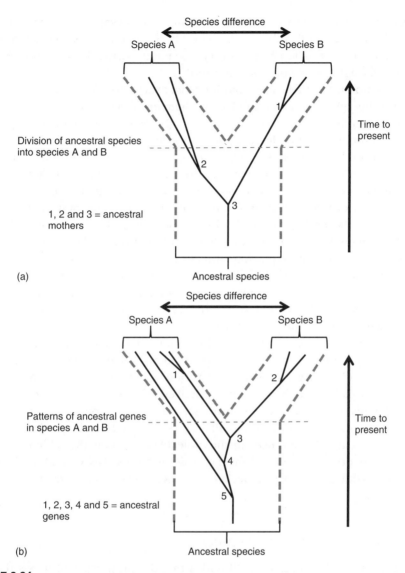

FIGURE 2.26

Frame (a) shows various possibilities for individuals (called 'mothers') in this example in a situation where an ancestral species has divided into two descendant species, A and B. For example, an ancestral mother B may be within the new species after the split (e.g. B). An ancestral mother may be within the original species before the split A. The ancestral mother of individuals in both A and B is highly unlikely to be at the split into A and B and may well be very much earlier in the initial ancestral species. Frame (b) shows the various possibilities for molecules (say genes) in a similar situation. Ancestral gene relationships can vary greatly from molecule to molecule. The trees of molecules may well vary enormously, and may not have any similarity to the tree of individuals. Again, different gene trees give relationships that vary enormously in time, including going very far back.

that humans (species A) were closer to chimpanzees (species B) than to gorillas (species C). It was gorillas that were the outliers. Such studies suggested that the human/chimpanzee relationship was very recent indeed, that is, only about 3 million years ago (though some workers demurred).

Yet the anatomies of these apes and humans, examined superficially, do not seem to concur; the apes are rather similar, the humans so very different. The implication was drawn that the anatomical data, though obviously containing considerable information about ancestor–descendant relationships, was so clouded by other information as to be less reliable for helping determine such relationships. This was, it was assumed, not true of the majority of the information in the biochemical data.

The 3 million year old recent common ancestor idea was, in turn, itself changed, in the last quarter of last century, in the reverse direction. Thus, new anatomical data resulting from the finding of much older fossils (judged to be on branches even closer to humans than australopithecines), and new molecular studies (especially various genetic similarities placing primate molecular relationships into a wider mammalian context) provided the new picture. These newer findings have pushed back the time of the human/chimpanzee common ancestor. Such times now include: 5–6 million years, 8–10 million years, and even 13 million years (and counting), depending upon study and investigator.

In other words, despite much more information, there is now much less certainty.

Anatomically there seems to be no doubt that humans have a more different anatomy from all the great apes than the various great apes do from one another. A great deal of this is to do, however, with derived adaptations to function in humans (such as bipedality) that evolved after the separation from apes, and to the retention of ancient adaptations to function in all apes (such as arboreality). When, however, we come to measure differences in the molecules, it seems that chimpanzees and humans really do share more of their genetic materials (DNA) (about 98.4%) than either does with gorillas (or especially yet more distant apes such as the Asian orangutans).

Anatomical Difference, Molecular Similarity

One explanation of this relates to the idea that the proportion of genes that code for obvious anatomical features is small (say well under 20%). This latter proportion is thus changed as a result of functional adaptation and gets swallowed up by the remaining portion of the genome (say 80% or more). The remaining portion codes for molecular, biochemical, cellular and physiological features (that are less well-known), as well as other less obvious anatomical features (such as patterns of blood vessels and nerves, complexities of muscles, bones and joints, details of skin and hair structure, and so on). These could well be held in common in any chimpanzee or human (or other ape and perhaps many monkeys). They imply that the 'rapid change in anatomy' concept for humans is not important in deciding relationships because it applies to only a very small portion of the total genome.

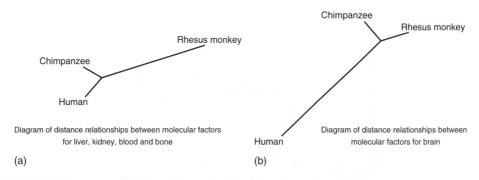

Diagram of distance relationships between molecular factors for liver, kidney, blood and bone

(a)

Diagram of distance relationships between molecular factors for brain

(b)

FIGURE 2.27

This is an example of real molecular data that follow the differences outlined in Fig. 2.24. In terms of molecular relationships in liver, kidney, blood and bone molecular factors, chimpanzees and humans form a closely related pair – it is rhesus monkeys that are far different. But in terms of brain molecules, it is chimpanzees and rhesus monkeys that are similar, and humans that are very different.

However, this argument has become more complicated in recent years as we examine various anatomical parts. For example, certain genetic factors may well relate to similar anatomies in, for example, chimpanzee and human livers, chimpanzee and human kidneys, chimpanzee and human blood, chimpanzee and human bones, and so on. Each of these pairs is far more similar than they are to the equivalent factors in the much more distantly related Old World monkeys. This fits with the idea of the close 'blood' (and liver and kidneys and bones) 'ties' of humans and chimpanzees Fig. 2.27(a)).

New information shows, however, that this is not true of molecular components of the brain. Chimpanzee brains are actually much more closely similar in their molecular factors to the brains of Old World monkeys. Human brains are far different in their molecular factors from both chimpanzee and monkey brains. In fact, the difference from chimpanzees (and monkeys) in these human brain factors is so great that it has been calculated that human brains must have evolved at from four to six times the rate of chimpanzee brains since the ancestral separation of pre-humans from pre-chimpanzees (Fig. 2.27(b)).

One might assume that this means that the genetic factors are the key; it is the anatomical factors that are confused. It turns out, however, that the molecular findings separating the human brain from the brains of chimpanzees and monkeys are mirrored by major aspects of the anatomies of brains in these species. When examined in ways that might be expected to reflect brain functions between major brain parts, we find that human brains differ anatomically from both chimpanzees and monkeys by a factor of almost 20 standard deviation units or more (Fig. 2.28). This places a measurement upon the separation of the human brain. And this is so even though we cannot take account of brain parts within the cortex (the limit to the analysis of Fig. 2.28) because intra-cortical parts of the brain (which are most likely

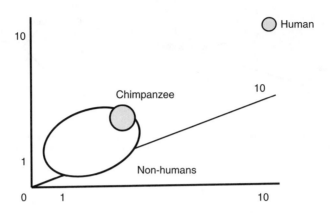

FIGURE 2.28

This is an example of real anatomical data on brain interrelationships that shows the same answer as the brain molecules (in Fig. 2.27(b)). In the three dimensions of the diagram (three multivariate statistical axes), human brains (the small circle at upper right) are some 20 standard deviation units different from all non-human primates (the large ellipse). Chimpanzees (the small ellipse) lie within the non-human primates. Thus, the human brain is enormously and uniquely different from chimpanzees as well as all other non-human primates despite the DNA similarity of chimpanzees to humans.

to be even more different from those of chimpanzees) could not be included in that study. All this implies that a new look is required at specifically brain evolutionary mechanisms in human evolution.

Clearly the evolution of the human brain is the greatest new thing that has occurred since the human chimpanzee common ancestor.

This entire discussion is not meant to imply that humans and chimpanzees as organisms are not very closely related – indeed they are. It is meant to convey, however, that the amount of difference between human and chimpanzee brains results from much greater brain changes in the lines leading to humans than in the lines leading to any apes. In contrast, though the length of time has been much greater from the much more ancient common ancestor of monkeys and chimpanzees, the pace of that change has been so slow that chimpanzee brains still far more closely resemble monkey brains than human brains.

It also means that it is not enough to look at similarities of chimpanzees and humans; we also have to look at differences.

The History of the Molecular Human/Ape Relationship

The coming of the 'molecular revolution' thus truly 'revolutionized' our understanding of biological evolution. The ideas started much earlier than we generally think. It is usual to credit workers like Crick and Watson (1953) and Zuckerkandl and Pauling (1975), and the molecular revolution of those times, with the idea that evolutionary relationships might be read directly through measures of molecular relationships. Yet such ideas were abroad even as early as the middle of the previous century.

For though Darwin knew nothing of modern biology, he wrote (Darwin 1859, p.485)

> Nevertheless all living things have much in common, in their chemical composition, their germinal vesicles, their cellular structures … and their liability to injurious influences.

How much closer could Darwin have come, given the vocabulary and concepts of his day, to the ideas of 'molecular, developmental, ultrastructural … and immunological' relationships of our times?

Only a little later, at the turn of Darwin's century, did the then newly discovered blood groups of Friedenthal in Germany (Friedenthal 1899) and Nuttall in England (Nuttall 1904), start to speak to chemical and molecular similarities in evolution. These studies were followed by a seminal review by Zuckerman (1933) that pinpointed very clearly the addition to anatomy, of physiological, biochemical, immunological and related phenomena for understanding the evolutionary relationships of primates.

This idea greatly flourished in the middle of the 20th century with the production of many studies such as immunology and amino-acid sequencing. This has continued in the last quarter of the last century from the results of a large number of methods and studies even closer to the hereditary materials (such as comparisons of restriction fragment length polymorphisms, and DND/DNA hybridizations). It has been taken very much further still with DNA sequencing.

Most remarkably of all, however, is today's knowledge of the total genome of a few exemplar organisms (including of course, humans). This will surely be extended to many more species very rapidly. Molecular evolution has now become incredibly important.

The Influence of Development Upon the Molecules

However, DNA similarity ideas are being further modified by new knowledge of the molecular mechanisms of development. Far from the blueprint DNA being the entire determinant of anatomical (and other) differences, the various developmental tool-kit molecules, their interactions with environmental factors and processes, and their effects upon brain/body interrelationships, are also being perceived as critically important. This may well be the next level of information about similarities and differences between humans and chimpanzees.

Thus, the effects of closely matching DNA blueprints (seemingly implying a great similarity) are, in fact, modified by molecular tool-kits (producing enormous differences). These in turn can be impacted by different environmental factors operating on the blueprints and tool-kits at different developmental times. All of these producing yet further differences. Effects on the brain and the brain–body interrelationship, acting uniquely in humans, are yet more complex. Understanding this new complexity may be the next level of information in the human story.

In other words, we need to be extremely cautious in assuming that we know the 'right' answers, a degree of caution that does not fit with media views of these matters (simplicities like 'missing links' and 'killer apes').

The Cladistics Approach to Humans, the 'Underlying Character'

All of this has been much impacted by the use of a technology called cladistics. In its theoretical development, cladistics is impeccable (Hennig, 1966). It involves the idea that, using mathematical methods, one can analyze the patterns of animal characters (of whatever kind, originally anatomical, but later all kinds of characters especially including molecular) in order to work out evolutionary relationships.

Thus, if groups of species share a set of characters (and especially if these include data on characters from species back in time, fossils) this set may be defined as 'primitive'. That is, they are present in these groups, because they were inherited from an initial common ancestor. If one group of species has characters that are different from those of its sister groups, then these are called 'derived' characters. These are new derivations in that group since it split from the line leading from one or more of the other groups. Part of these definitions of the characters requires comparisons with other fairly closely related groups that nevertheless do not belong to the sets of groups being examined. These are called 'out-groups'. As I said, the theory (Hennig, 1966) is impeccable.

However, it is immediately obvious that the decision that cladistics requires, that characters be assessed as either primitive or derived, involves special practical matters.

One of these matters is that similar characters may not be due to common inheritance (primitive), but be obtained in parallel (derived). Similar characters may even be similar because, originally different in two groups, they changed during the course of evolution towards becoming similar in the two groups (converged). Parallel and convergent evolutionary patterns are well known situations. The methods of cladistics can handle these situations.

Another matter is that, for such analyses to work, it must be possible to make the dichotomous decision that a given character is 'primitive' or 'derived'. It is at this point that we realize that it is probably rather rare, especially for anatomical information, for a character to be specifically singularly determined. There are good examples of singular determinations: eye color, for instance, may be specifically related to an individual gene configuration. However, such single determinations are almost certainly not common for most anatomical features. Much more frequently a given character will be related to a wide packet of genes. This is a major difficulty for cladistics.

The Cladistics Approach to Humans, the 'Observable Feature'

Much the greatest problem for cladistics, however, is that what has been called a 'character' in the foregoing discussion is really only what one observes (or measures). It is actually an 'observable feature'. The great majority of observable features will, themselves, be some compound or resultant of several different underlying features, each of these being determined by different genetic mechanisms. And these in turn may depend upon yet a finer array of true developmental determining mechanisms. In such cases, it would be unlikely that the entire group of such underlying observable features would be primitive or derived. It is much more likely that the original observable features will be some compound of underlying observable features, at several different levels, and with some being primitive and some derived.

A practical example may make this clearer. One anatomical character that is used in studies of humans (and other primates) is simple overall relative skull length. A relatively shorter skull is said to be 'derived' in humans as compared with a relatively longer skull ('primitive') in apes!

But overall skull length is a compound of the lengths of three parts of the skull, the anterior, middle and posterior cranial fossae. To further complicate matters, each of these three subunits of skull length are themselves associated with further quantities: different thicknesses of the cranial bones, different sizes of air sinuses, different heavinesses of cranial buttresses, different vascularities of skull bone marrows, different patterns of skull bone sutures, just to enumerate a few (Fig. 2.29). Whatever the determinants of all these factors, they are extremely unlikely to be all primitive or derived. In other words, a singular cladistic determination for skull length as primitive or derived, whichever it may be, can only be wrong.

Further, just assuming that these many individual mini-lengths lying within that original skull maxi-length might add up to overall skull length is also naive. There must be many and varied ways in which these sub-measurement are integrated in producing the final 'overall skull length'. Of course, this is reminiscent of fractals; one can continuously break up such maxi-features into milli-features, micro-features, nano-features, pico-features, and so on.

As a result, when the decision is made in practical cladistics that a 'character' is primitive or derived, about the only thing of which we can be fairly certain is that either determination, whatever it may be, will be wrong. The real determination, in most cases, will be that the 'observable feature' (whether meristic or quantitative) is the resultant of a complex of underlying mechanisms, some proportion of which will be primitive, some other proportion derived, and possibly yet other proportions interacting, perhaps even in several different ways, and how do we know?

This might give rise to the idea that one could partition the contributions between primitive and derived. Except that even that idea is likely to be more complex. It is entirely likely that a simple partitioning, the idea that one part can be simply added to the other part, will be wrong, because interactions among the various hereditary mechanisms, developmental processes, and environmental factors may result in multiplicative, not additive, relationships.

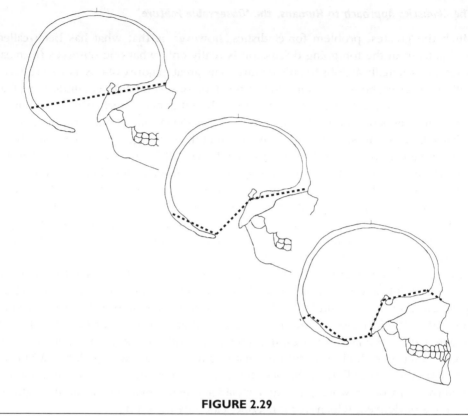

FIGURE 2.29

Three ways of viewing what makes up cranial length in humans.

Some observable features, clearly genetically determined, may change in a linear fashion due to selection within a linearly changing environment, for example, increasing climactic dryness, increasing deforestation, increasing geographic separation. Other observable features, perhaps equally genetically determined, may have been influenced by selection within cyclically changing situations (e.g. change from tree-living to ground-living to tree-living, or change from ground to swamp to ground, and for each of, possibly, a succession of such cycles). Yet other observable features, also genetically determined, may have been influenced by quite complex and absolutely non-linear, non-stochastic processes, in terms of biomechanical, physiological, biochemical or other such adaptations of the whole organism.

One might think that these complications of anatomy would be much less likely to be the case for molecular 'characters'. After all, so the argument might go, molecular characters are, genuinely, so much closer to the primary genetic mechanisms. However, modern developmental biology shows that even here the situation is much more complex. There are many developmental mechanisms and environmental factors interpolated between initial individual molecular features and their eventual hereditary determinants. There are many similar DNA components that give totally different anatomies depending upon the complex course of development and the complexities of epigenetic interactions. And what is more, these complexities are multiplying as developmental biology moves into its new century.

2.6.3 Some Further Thoughts on Evolution and Human Anatomy

Darwin's "On the Origin of Species" was very careful about the place of humans among the primates. Darwin, originally and truly, but only briefly, recognized the similarities of humans and apes. It was left until later years and later workers to speculate just what this similarity really meant.

A result of this was that, as a reaction to Darwin's ideas, many efforts were made to show how humans were quite distinct from apes. The most famous of these was perpetrated by Darwin's most obvious opponent, Owen, who put forward the idea that only humans possessed a "hippocampus minor" a small area of the brain. This was, of course, a major win for the Darwinian side (seized upon by Huxley (1863) in open debate) because it was so easy to show that it was wrong.

'Ever since Darwin' (famous phrase) the anatomical differences between humans and chimpanzees have been whittled down. Anatomists have been dissecting away all those bits of the human body that are common to chimpanzees and humans. The bones – they can go – essentially the same even if a bit 'contorted' in chimpanzees. The heart – that can go too – same in chimps and humans. The brain, even that can go – 96.6 of brain size in both chimps and humans is simply the result of differential body size! The muscles can go – same in humans and chimpanzees except for one or two little muscles, one of which enjoys the sonorous title of 'dorsoepitrochlearis'. Yet only last year I found that, though the existence of that little muscle is known in some 5% of humans, its connective tissue 'ghost' exists in every human.

Likewise, ever since Darwin, behavioral scientists have been chipping away at what we do. 'Man' (sic) was originally different. Only Man (sic) uses tools – no – lots of animals use tools. Aha, only Man (sic) makes tools – no – a number of animals make tools, even some very surprising ones. Aha, aha, only Man (sic) makes tools for the future – no – some animals even do this. Aha, aha, aha, only Man (sic) speaks – well perhaps, no – though not actually talking, many animals communicate very well. There seems to be almost nothing behavioral that separates us, except in degree (and some would have it only a very minor degree) from chimpanzees and other folk.

Thus, ever since Darwin, Man (sic) has been continually demoted. He fell from the zoological superfamily named after him (the Hominoidea) to his own family (the Hominidae), then down again to his own subfamily (the Homininae). All the way down he has been followed by the chimpanzee and gorilla. It reminds me of the cartoon of the senator from Kansas, who, looking over his shoulder and noting the fossil characters in a line behind him, says: "Will you quit following me". Humans are currently in their own genus (*Homo*). While sharing that with fossil species (e.g. *Homo erectus*) we are now under some pressure to share even with the living chimpanzee and gorilla. Even our place as a species (*Homo sapiens*) has been questioned; there was a day when we were relegated to mere subspecies status, *Homo sapiens sapiens*, standing along-side *Homo sapiens neanderthalensis*. Then we were separate. But the new DNA information about Neanderthals, the possibility of humans sharing some Neanderthal genes may yet put us together again. Today things are changing in reverse.

When I was a child, I thought, as a child, that the common ancestor of modern humans and modern apes was some creature perhaps 20, 25 or even 30 million years ago. A respectable degree of distance! One that the 'Bishop's Wife' could tolerate!

When I became a man I put away childish things. New fossil anatomies, new animal behaviors, new concepts of molecules, and new views of brains, were all drawing the human/chimpanzee relationship ever closer – eventually indecently close! Thus, by the middle of my life, the human–chimpanzee link had become reduced to just 3 million years (some said even less)! The culmination of this trend was the oft-quoted DNA fact that humans are 98.6% chimpanzees. Of course, there are other, rather less quoted, facts that humans are 96% orangutans, 90% rats, and even 55% bananas.

As I grow old, however, exciting things are happening. New alliances, new integrations, of anatomies, of developments, of brains, of behaviors, of life histories, and of environments, with evolution are starting to imply something more. Humans are actually far more different from chimpanzees than we thought. Of course, human livers, kidneys, spleens, muscles, bones, and so on, are all extremely similar to chimpanzee livers, kidneys, spleens, muscles, bones, and so on. Both of these are vastly different from monkey livers, kidneys, and so on.

But when it comes to brains, the similarity is between chimpanzees and Old World monkeys. In quantitative terms chimpanzee brains are only 3 standard deviation units different from rhesus monkey brains. Human brains are an incredible 30 standard deviation units different from both chimpanzee and rhesus monkey brains. It is humans that are vastly different.

2.6.4 A New Evolutionary Mechanism

One can envisage that a new form of evolution is now operating in humans. It is one not available to animals. It is one produced by modifications during development from new interactions between genes more ancient than we had ever understood, newly understood developmental cascades, even newer environmental interactions, further newly recognized influences up and down the human generations, totally new interactions across and between human communities, and absolutely new physical (world travel) and electronic (world communication) exchanges across the globe. None of this is available to any animal.

They are resulting in enormous changes, and enormously increasing rates of changes, of humans from any other creature. And they are, perhaps, more evident in the human brain and brain function than in any other human component. Far from human evolution slowly drawing to a close, as some have thought, a new human evolution is opening up in amazing ways not evident to even a very percipient Darwin.

H. G. Wells (1933) looked forward to *Homo sapiens* (Man the wise) becoming *Homo superbus* (Man the superb) the last achievement (as he thought) of evolution. Today, I have one colleague (Professor Alan Harvey who, with the aid of a classical scholar) is elevating the *Homo sapiens* (Man the Wise) of perhaps thirty thousand years ago, to the *Homo sapientior* (Man the Wiser) of today. And for the future: it is possible that we might see modern humans actually split (not into the Morlocks and Eloi of the time machine, Wells, again, but 1895) but into *Homo sapientissimus* (Man the Wisest) and *Homo nerdensis* (Man the Nerd) as we pass from the Fishing Net to the Inter-Net.

How creation scientists will **love** some of these new findings.

How exciting this all is. If only I were at the beginning rather than at the end.

CHAPTER THREE

'The Naming of the Parts':
Some Wrinkles

The term 'wrinkles' can, of course, really just mean wrinkles. Wrinkles are ubiquitous in humans, from wrinkles in the newborn baby, wrinkles of hands immersed in water, wrinkles of the face in pain or in depression, wrinkles of dehydration, wrinkles following pregnancy, wrinkles of extreme weight loss, wrinkles of age, and so on. Of course, these wrinkles have a variety of causations. Some appear rather opposite: from excess of watery fluids (e.g. newborns newly departed from the watery amniotic environment, and hands in washing-up water) to lack of fluids (dehydration, fluid lack or fluid loss); from increased activity such as facial muscle contractions (e.g. in pain or depression) to loss of muscle activity (e.g. following weight loss, pregnancy, and with aging),

The Scientific Bases of Human Anatomy, First Edition. Charles Oxnard.
© 2015 John Wiley & Sons, Inc. Published 2015 by John Wiley & Sons, Inc.

from increased facial wrinkling (e.g. of progeria) to decreased facial wrinkling (e.g. of myxedema), and so on. All of these are important wrinkles in human anatomy.

The word, 'wrinkles' also has another meaning. Wrinkles are tips that help us to avoid confusions or see our way through difficulties. It is to this concept that this chapter is devoted. Among the first things that beginning health science professionals need to get straight in their minds are the descriptive terms in relation to the human body. Human medical anatomy text books almost always define these descriptive terms in early chapters. They become (and must become) absolutely second nature to every health professional. No one afterwards has to think about them. They must become automatic.

Not only are they necessary to allow health professionals to talk to one another with certainty, they are also essential so that no one removes the wrong kidney, operates on the wrong limb, even on the wrong patient. Yet for all that, mistakes occur. Hence, in the medical world, notwithstanding the fact that these matters are emblazoned on the mind, precise procedures are undertaken to attempt to prevent such mistakes. Even so, human nature being what it is, mistakes occur (and probably will continue to occur) however much we attempt to tighten procedures by using 'world's best practise'!

In a similar way, descriptive terms are also necessary for scientific understanding. This has been especially emphasized in the 'birds-eye view' that we have just examined. However, because the birds-eye view has to ally adult human anatomy with the anatomies of development, the anatomies of other creatures, and the anatomies of function, the bird's eye view uses different descriptive terms. These terms are more than just descriptive; they have specifically arisen to allow the scientist to understand the science behind the terminology. But the medical terminology is special to humans, and often takes little or no account of the science. A discussion of human anatomical descriptive terms is therefore needed in order to aid understanding of the science underlying human anatomy.

I therefore do not apologize for drawing to your attention differences in terminology between human topographic anatomy on the one hand, and the scientific anatomies of development, comparison and function, on the other. If not understood, these differences can be the sources of confusion.

3.1 Terminological Confusions

These confusions are due to different terms being applied in adult humans, as compared with developing, evolving and functioning individuals, to: the position of the body, the placement of body parts, the relationships of body structures, and changes in all these during development, function and evolution. Let us look at this in more detail.

Descriptive terms are applied in the transition from zygote to embryo. Changes then occur in the zygote–embryo–fetus transition that affects the terms. Further changes then occur in the baby–child–adult transition that follows. These changes relate to the developmental changes that are part of the science underlying human anatomy.

Other descriptive terms arise from the different postures and positions that characterize many animal anatomies. They are due to differences of form in creatures that swim, in creatures that move on four limbs, and that creature, human, that is upright, moving bipedally.

Finally, yet further terms arise according to the different positions: standing, sitting or lying, in which the human body may be described and examined.

Thus, the embryo–fetus transitions are usually figured with the convex surface (the back) at the top of the page and the concave part (the belly) at the bottom of the page (especially when looking at transverse sections). This is certainly the convention that has been followed in the bird's-eye views of the last chapter. The bulbous part (the head or crown) is usually figured towards the left of the page, the pointed part (the tail, or breech) towards the right (especially if we are looking at longitudinal views). The terms that apply to these four directions are dorsal and ventral, cranial and caudal (merely the Latin words for these English names in brackets), respectively.

However, in the baby–adult transitions (the baby usually figured lying, the adult in the standing position), parts that were described as cranial (towards the head) in the fetus are often described as superior (above) in the adult, parts that were caudal (towards the tail) in the fetus, as inferior (below) in the adult, parts that were dorsal (back) in the fetus as posterior (behind) in the adult, and parts that were ventral (belly) in the fetus, as anterior (front) in the adult.

In the animal comparisons there are further clashes. In animals, though cranial is the correct word for things near the head end, they are also called anterior, because they go at the front (cf. anterior in the previous paragraph). This comparative usage is identical to the developmental use. Thus, the 'developmental' and 'comparative' usages of anterior are at ninety degrees to the human usage. Or, if you like, the human usage of anterior is at ninety degrees to the developmental and comparative. Confusions like this apply to many of the other medical terms.

Even in regard to function, confusions can apply. Thus, the muscles on the front of the thigh are called anterior in medicine (because they are in the front). But the equivalent muscles in the arm are posterior in medicine (because they are on the back of the arm). From the viewpoint of function, however, both sets of muscles are dorsal because they are supplied by dorsal nerves, originated on the dorsal aspect of the limb, and participate in extensor functions and reflexes, and so on.

Of course, from a historical perspective, the human terms came first. But initially they were 'just' descriptive; the science behind them had not been clearly worked out. As a result they hide the developmental, evolutionary and functional underpinnings of the human anatomy. To really understand the science behind human anatomy it is best for us to use terms that are similar in the transitions from embryos to adults and in the comparisons across animals to humans. It is important to understand this in the move from overall views of anatomies (the prior chapters) to specific descriptions of human anatomies (the following chapters). Let us examine it in more detail.

3.2   Implications for Names from Developmental Anatomy

Thus, as the comma-shaped embryo mutates into the fetus and baby, various new structures are formed within it. The central nervous system develops close to the back and is therefore dorsal in position. The notochord develops in the center and is therefore axial in position. The gut develops close to the putative belly and is therefore ventral in position. That seems simple.

However, during these developmental transitions, many structures that originate in one position end up in other positions. These are due to a variety of mechanisms. Some involved divisions of structures, some migrations, and some rotations.

3.2.1   The Effects, on Names, of Developmental Divisions

Take the mechanism of developmental divisions of structures (we have already examined this in Chapter 2). The precursors of the trunk musculature become divided into dorsal and ventral blocks that originate in relation to the central (axial) notochord. The dorsal muscle precursors originate in a dorsal position near the dorsal nerve cord. The ventral muscle precursors arise alongside the more ventrally placed gut tube. Naming these muscle precursors as dorsal and ventral in the trunk is not just descriptive, but fundamental, directed by specific developmental factors.

But in the adult human, the ventral muscle derivatives are so much larger than the dorsal that many of their eventual components actually come to be positioned very dorsally in the trunk. For example, one such component will become the quadratus lumborum muscle. This is a muscle that, in medical terms, lies at the back of the abdomen, on the posterior (medical term) abdominal wall, and is therefore dorsal in position. But it is a developmental derivative of the ventral muscle block. If we did not understand that it is 'really' ventral in derivation, then we would puzzle as to why this 'dorsal' muscle is innervated by ventral nerve branches, and supplied by ventral blood vessels, that is, is really ventral in origin. In other words, knowing the underlying mechanism is the science of why things are where they are. Let us follow this further.

We have already learnt that a second order division occurs within the limbs. Of course, limbs develop only from the ventral compartment of the trunk. Therefore limb muscles are all ventral components of the trunk that have migrated into the limb. They are innervated, therefore, by ventral branches of nerve trunks passing into the limb, and supplied by ventral vascular components. But there is a second order division of these ventral components into dorsal and ventral in the limb itself (a solely ventral structure). This new division applies to the skin surfaces and the differentiating limb muscles within the limb; they have their own dorsal and ventral organization; and the trunk ventral nerves and vessels supplying the limb likewise become divided into dorsal and ventral within the limb.

Understanding this allows us to work out why particular muscles are supplied by particular nerves. Memory is replaced by understanding.

This becomes further complicated because some limb muscles migrate during development. When they do so, they may come to lie in a different position. Let us examine this further.

3.2.2 The Effects, on Names, of Developmental Migrations

In limbs many of the most proximal muscles (proximal meaning nearer the center) that are important in relation to the functions of limbs, migrate (during development and/or evolution) even more proximally into the trunk during development. Latissimus dorsi is an example here. Though a secondary dorsal muscle block within the limb, it derives, as do all limb muscles, from the ventral division of trunk musculature. But in the migration back into the trunk, its proximal attachments pass from the limb back onto the dorsum of the trunk. Indeed, on the trunk it lies not only totally dorsally (as is obvious from its name – musculus latissimus dorsi – the broad muscle of the back) but it even covers over the true dorsal trunk muscles (the erector spinae – the muscles of the vertebral column). The possible confusion is obvious. This muscle is an embryological ventral muscle; it is dorsal in position in the adult. The fact of its real embryological derivation is seen from the fact that it is supplied by trunk ventral nerve branches. Knowing this developmental history provides understanding of the adult situation.

There are many other examples of such developmental migrations. One very obvious one, noted earlier, is the migration of the diaphragm precursor from a very cranial position (cranial even to the precursor of the brain) in the early embryonic plate into its eventual position at the thoraco-abdominal boundary in the fetus (and baby and adult). As a result, its nerve supply does not come from the spinal segments at the thoraco-abdominal boundary (perhaps lowest thoracic spinal nerves), but from those upper cervical nerves received as it passed the neck in its caudal descent from, originally, a position even more cranial than the brain. Thus we can understand why the diaphragm is supplied by nerves from cervical segments (C3, 4 and 5) and why the phrenic nerve (from these cervical segments) supplying the diaphragm has such a long pathway from C3, C4 and C5 in the neck caudally through the thorax to the diaphragm at the thoraco-abdominal boundary.

Yet another similar example is the heart. It originates as a series of small vessels that lie in the embryo far above, that is, cranial, to the developing head, with the aforementioned diaphragm precursor even more cranially. During development, however, the ventral folding (head fold), already mentioned for the diaphragm, also involves the developing heart so that it too moves ventrally and caudally in front of the developing brain. This fold continues during early development so that, sequentially, the heart rudiments pass caudally and ventrally, past the head, past the neck, and end up as the heart lying in the thorax. Understanding this process is critical to understanding why the vagus nerve (coming from certain brain segments in the head) also supplies the thoracically located heart. Yet again, knowing that this migration has occurred provides key understanding of the adult situation.

The final set of examples involves the nerves in the trunk that derive from each of the spinal segments.

The nerve supply of the skin of the trunk comes from, obviously, both primary dorsal and ventral branches of the nerves (spinal nerves) to each body segment. The ventral skin areas, like the ventral muscles, are more extensive than the dorsal. The ventral nerve branches, therefore, are bigger than the dorsal, and innervate a much larger portion of the trunk skin. They supply not only all of the truly ventral skin, but also all the lateral skin, and most of the dorsal skin. Only the central dorsal strip of skin is supplied by the dorsal branches of the spinal nerves. To think that the whole of the dorsal skin should be supplied by dorsal nerves alone is to misunderstand the developmental science behind cutaneous nerve supplies.

A second example can be seen in the nerve supply of the limbs. Thus, in the trunk generally, there are (overlapping) skin areas that are supplied by segments sequentially along the trunk. That is, in the middle of the trunk, the spinal nerves, shall we say, of the 7th, 8th, and 9th thoracic segments supply skin sequentially in overlapping bands around the trunk. For example, when there is a medical problem involving nerve distributions in the trunk (e.g. herpes zoster) the blisters (and the pain) are arrayed in a band around the trunk in the distribution of that particular spinal nerve. However, at the point where the limbs are attached to the trunk, the developmental material that is drawn from the trunk into the limb during development is very extensive, covering several segments. As a result, the segmental nerve supplies are also drawn into the trunk with the result that there is actually a deficit in the trunk nerve supplies. Thus in the neck the segments are supplied from the various cervical nerves, in the lower thorax from the various thoracic segments, as explained above. But in the upper thorax, in the region of the upper limb, the nerve supply to the ventral surface of the trunk passes from the 4th cervical to the 1st thoracic. The intervening neural segments are drawn almost totally into the limb. A similar phenomenon occurs in the lower abdomen in relation to the lower limb. These gaps in segmental supplies of the trunk do not occur on the dorsum; why would they; the limbs are ventral trunk derivatives! Once again, therefore, the science explains what would otherwise require memory.

3.2.3 The Effects, on Names, of Developmental Rotations

In the early fetus, the limbs are lateral projections from the trunk and, though totally ventral trunk derivatives, they have their own internal dorsal and ventral sides. However, during development, the limbs, both fore and hind, become rotated medially. As a result, thus the thumb and the big toe, which were first both placed on the original cranial border of the early limb bud, come to lie medially on the rotated limb. Similarly the original skin surface that faced dorsally becomes twisted around so that it is now facing ventrally. This explains why the ventral skin in the lower limb is innervated and supplied by dorsal nerves and blood vessels (and vice versa). All of these structures thus twist around the baby limb with the twist of the muscle blocks. Knowing that this developmental twist has occurred provides understanding of the adult situation.

However, the aforementioned is not the whole story. The upper limb is further modified. Thus, though the baby's thumb and big toe are medial in position on the limbs, in the adult, the thumb, but not the big toe, is described in medical anatomy as lateral in position. This is because, in their wisdom, the older human anatomists decided that the 'anatomical position' was one in which the forearm (and its bones, radius and ulna) were placed so that the hands were supine (lying on their backs). The anatomists could not do this with the foot because, in humans, there is no pronation and supination allowed between the two leg bones, the tibia and fibula. The result of this is that, as one looks at forearm structures in adults in the 'anatomical position' there is a secondary and opposite twist of the forearm that places the thumbs laterally. This secondary (and actually artificial, not developmental) twist also affects the positions of the vessels and nerves of the forearm. Yet again, knowing that this further twist has occurred provides understanding of the real adult situation. We will go into more detail about these limb rotations later.

3.2.4 The Effects, on Names, of Function

Finally, because these concepts apply to structures, they also apply to the functions of structures. For example, the aforementioned division of the limb musculature into dorsal and ventral muscle blocks means that the movements that these muscles can produce at their respective joints are dorsalward and ventralward movements. These movements are also called extension and flexion, respectively. But when these muscles become rotated, the movements that they produce are changed.

As an example, look at the muscles on the top of the foot – pointing in the same direction as the belly. These muscles are not ventral muscles. They are embryologically dorsal muscles that, because of the rotation, lie on what a patient would call the front of the foot. These embryologically dorsal muscles have dorsal innervations, dorsal vascular supplies, and dorsal movements; that is, the movements that they produce, drawing the foot and toes upwards, are really dorsal movements, and therefore should be called extension. Because of possible ambiguity (the patient would call the opposite movement extension) this movement is called 'dorsi-flexion'. It is to be distinguished from the reverse movement, which is embryologically flexion, but again to avoid confusion, because the layman calls this movement extension – as in extension of the lower limb by a ballet dancer – it is labeled 'plantar flexion'.

That these movements are truly embryological is evident from the functional reflexes that go with these muscles. Thus, pressing on the sole of the baby's foot reflexly makes the lower limb straighten (extend) but the foot move into the position of standing (dorsiflexed). This is the 'standing reflex that allows a very young baby to appear to be standing. The opposite occurs when a noxious stimulus (perhaps a gentle pinch) is applied to the foot. The babies entire lower limb is withdrawn (flexion of hips and knees), but the foot is plantar flexed – that is, pointed. This is biological flexion (clearly seen in the opposite nervous reflex, the nociceptive reflex, the reaction to a noxious stimulus).

3.2.5 Changes, in Names, from Comparative Anatomy

The developmental and functional examples given above are useful exactly because they are the result of the changes that occur, from the embryo to the adult, in every single human. If, however, we compare the adults of animals living today, we can see a series of what look like similar processes.

One example involves the evolution of limbs. During the evolution from fins in fish-like ancestors to limbs in tetrapod ancestors, many changes have occurred in the positions of the parts. These are well described in many comparative anatomical texts and they relate to changes in function. They almost certainly include many separate parallel sets of changes, in the long (and, I believe, likely several separate) transitions from water to land. However, among those changes are a series that are described as rotations of the limbs. Thus forelimbs are swung one way and hindlimbs the other, so that, eventually, elbows come to point backwards and knees forwards. The complex evolutionary process described here is not the same as the rotation of the limbs that occurs in human development. It only looks somewhat similar. It is easy to assume, wrongly, that this is also the 'explanation' of the developmental phenomenon.

It is worth giving another, in this case, hoary, old example. Thus the gill slits of fish were said to explain the branchial arches of the human embryo. There is a major truth hidden in this incorrect generalization, but that was not what was originally meant. In fact, branchial structures in fish embryos give rise to gill slits in fish adults with respiratory functions. Branchial arches in human embryos give rise to many structures that have completely different adult functions. The idea that branchial arches are part of the blueprint of both fishes and humans is correct. But though these blueprints have elements of similarity in each species separately, they actually show marked differences to one another as they develop in the various creatures. The one never did give **rise** to the other. It is easy to read the underlying science incorrectly.

It was recognizing such processes as these that led into one of the blind alleys of evolution – the idea that development (ontogeny) recaptures evolution (phylogeny). This was often expressed in comparative terms as the later stages of earlier forms being the **reason** for similarities with the earlier stages of later forms. Notwithstanding this problem, such comparisons do provide useful information about names.

Yet these examples are the reverse of what was discussed earlier. They give rise to features that look as though they were related, but are not. They have, indeed, often been offered as 'explanations' of structural arrangements in humans when in fact they are spurious.

3.2.6 Types of Anatomical Names

In different organisms, corresponding parts can, therefore, be considered as similar by three criteria: ancestry, function, or appearance. The term homology (adjective homologous) applies to two or more features that are thought to share a common ancestry – though full direct evidence is usually lacking. The term analogy applies to two or more features with a similar function. Of course, it is possible for a given comparison to be both homologous and analogous. Structures that look alike (but whose functions and origins may be different) can be described as showing homoplasy (but I will not use this term further in this book).

A special case of homology is serial homology: developmental similarities between successively repeated parts in the **same** organism. The chain of vertebrae in the spinal column, the several branchial arches, and the successive muscle segments along the trunk, are examples.

These phenomena are, each, separate contributors to the design of organisms. They also mean that further terms can be applied to the form of organisms.

Thus, symmetry describes the way in which an animal body meets the surrounding environment. Radial symmetry refers to a body that is laid out more or less equally around a central axis so that any of several planes passing through the center divide the animal into several more or less equal or mirrored sectors. This is evident in invertebrates such as starfish and sea urchins. Humans (as with most other vertebrates) are bilaterally symmetrical, meaning that only a single plane divides the body into two mirrored halves: left and right. Even so this is an underlying concept rather than a complete reality; many aspects of the body are not mirrored from left to right – the position of the heart, the structures of the lungs, the arrangements of the gut, and so on.

Equally, segmentation is evident when a body or structure is built of repeated or duplicated sections. Yet again, duplication is the concept rather than the reality; such segments are not totally duplicated; variations in segments account for an enormous amount of anatomical structure. The various vertebrae are not totally duplicated. Variations from region to region occur, for example neck, thorax, and abdomen. It is easier (developmentally) to alter old parts than to add new parts.

3.3 Conclusion

Such scientific complexities extend to names in almost every system of the body. Not understanding them means that anatomical facts are feats of memory. Understanding them means that anatomical facts are derivative; they can be worked out. Understandings are absolutely necessary as we pass from 'the bird's-eye view' of human structure exemplified in the previous chapter, to the 'detailed aspects' of human structure described in the rest of this book.

CHAPTER FOUR

Building the Human Trunk

4.1   The External Trunk: From Plan to Layout

In Chapter 2 it was shown, on a variety of grounds, developmental, comparative, functional, integrative and evolutionary, that it is not unreasonable to consider the body in three parts: the trunk, the head and their respective 'brains'. The 'head and its brain' is extremely complex so as the 'trunk and its brain' (the spinal cord) is somewhat simpler, it is easier to consider it first. The trunk itself, as also shown in Chapter 2, can be considered in two parts: the external trunk and the internal trunk. Furthermore, because the trunk is largely completed in the transition from embryo to late fetus, we can carry our understanding directly on to the adult.

The Scientific Bases of Human Anatomy, First Edition. Charles Oxnard.
© 2015 John Wiley & Sons, Inc. Published 2015 by John Wiley & Sons, Inc.

In considering the changes in the external trunk from the embryo through the fetus to the adult, it may be useful first to understand the building materials, and second to recognize the building processes that are applied to them.

In Chapter 2, the developments were largely described as a repetitition of units along the length of the trunk. Gradually, during growth and development, further changes occur that mean that the various units (segments) become differently identified and modified. We can see, at a very early stage, a division of the mesoderm into a medially located **paraxial mesoderm** and a laterally located **lateral plate** mesoderm (Fig. 4.1(a)).

The first component, the paraxial mesoderm, eventually to leads to much of the external trunk, comes to lie dorsally as the embryo folds, and becomes segmented along the length of the trunk (Fig. 4.1 (b)).

The second component of the mesoderm, the lateral plate mesoderm, eventually leads to many components of the internal trunk. It is not segmented (in the trunk). It comes to lie ventrally as the embryo folds, and becomes arranged as deep and superficial layers (Fig. 4.2 (a,b)). Intermediate and at the apex between these layers in a dorsal location is a small component known, therefore, as the intermediate mesoderm. The intermediate mesoderm does exhibit a few early longitudinal segments but these soon disappear. These lateral plate mesodermal components relate to the developing internal trunk and are discussed later.

4.1.1 The Basic Components of the External Trunk: Myotome, Sclerotome, Dermatome

We already know that the somites of the mesoderm, through their differentiation into myotome, dermatome and sclerotome at each segmental level, provide the basic materials that will eventually become the adult body wall. The somites initially look similar but various developmental factors modulate their form along the trunk so that different portions (regions) of the trunk come to be readily identifiable.

The cranial-most of these regions includes overlaps with the head as the back of the skull. Its segments are called occipital somites; there is more than one of these but they are not readily identifiable and therefore not certainly numbered. These will not be treated here but will be handled when we come to study the head.

A second region becomes the neck, with segments (somites) that are labeled Cervical 1, Cervical 2, C3, C4 … and hence to C8 in humans. A third is the thorax with thoracic segments similarly labeled T1 to T12, a fourth the abdomen with lumbar segments labeled L1 to L5, and a fifth the sacrum with sacral segments labeled S1 to S5. There is also a sixth, the tail, or coccyx in humans. These caudal segments may be very numerous in creatures with long tails; but in humans they are very small, are not evident externally and not usually numbered.

At each segmental level these somites become divided into dorsal and ventral parts by molecular factors (under control of Hox genes) that are secreted by neighbouring structures. Thus the more dorsal parts of the somites lie close to the developing nervous system and are influenced by molecular factors (e.g. sonic hedgehog) that emanate from the nervous system. The more ventral parts of the somites lie close to

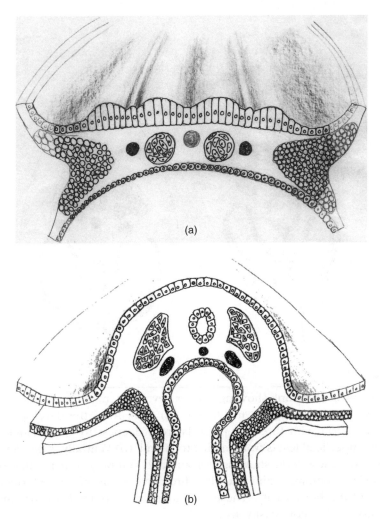

FIGURE 4.1

Cross-sections of early trunk showing: (a) the midline structures from dorsal to ventral: neural tube, notochord and gut tube, and (b) the bilateral structures from dorsal to ventral: somatic mesoderm (in two parts), intermediate mesoderm, and lateral plate mesoderm in two parts around the body cavity. These diagrams are 'exploded' so that the various components can be more easily seen.

the notochord and are influenced by molecular factors (e.g. also sonic hedgehog) that emanate from the notochord. This dorsoventral division affects all three regions of the somite: the myotome (a precursor of muscle covered by deep fascia), the sclerotome (a precursor of connective tissues, especially cartilage and bone, and

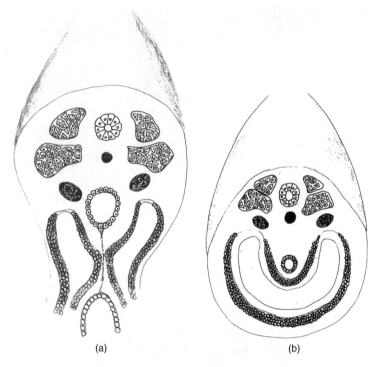

(a)   (b)

FIGURE 4.2

Cross-sections of later trunk showing further development of the somatic mesoderm: dorsal mesoderm showing the myotomes. (a) Dorsally is the development of the epaxial mesoderm with three deep levels from deepest to more superficial, and three more superficial levels from medial to lateral. (b) Ventrally is the development of the hypaxial mesoderm showing dorsally and ventrally near midline myotomal blocks as single columns, and laterally, three layers, superficial, intermediate and deep around the body cavity. These diagrams are 'exploded' so that the various components can be more easily seen.

their surrounding deep fasciae), and the dermatome (a precursor of the dermis including its superficial fascia underlying the cutaneous epithelium).

4.1.2 The Fate of the Myotomes

The myotomal blocks in the trunk give rise to the true trunk skeletal muscles lying under the deep fascia (see later). Those limb muscles that attach to the trunk, and special superficial muscles that lie in the superficial fascia outside the deep fascia, are considered later.

The first elements in the myotome of the somite are cellular precursors of muscle cells. These fuse into multinucleated myotubes, and these in turn become differentiated into muscle fibers. This is a process under control of a cascade of complex molecular factors. The muscle fibers themselves then become organized into blocks that eventually become the anatomically identifiable muscles of the adult. Non-muscular elements of muscles (internal connective tissues, tendons) come from different molecular factors that produce connective tissue differentiation, affecting the sclerotome (see later). It is, thus, possible to have muscles without their connective tissue elements, and, what happens more frequently, connective tissue elements, extra muscular fascial sheets, intra-muscular connective tissue sheets, and even tendons, all without muscles.

The muscle precursors first separate in a fundamental division into dorsal and ventral precursor blocks. At first they present as a pair (on each side) of individual segmental blocks at each level, they differentiate into layers within each block. The division of the myotomal blocks into layers is usually in relation to depth within the segment, that is, the layers are from deep to superficial. The deepest layers usually represent only one segment (or perhaps two) crossing between one (or perhaps two) vertebral precursors. The more superficial layers usually involve several precursor blocks covering several segments. It is from these layers that the various individual muscles crystallize out, as it were.

The actual arrangement of the muscles forming from these layers seems to depend upon the particular animals in which the process is examined. For example, in some animals, a particular myotomal layer may comprise a single muscle sheet spread over many segments. This is common in amphibians. In others, the single sheet may differentiate as a series of many similarly organized muscles over a sequence of segments (frequently so in reptiles). In yet other species (and this is especially so in humans) intervening portions of such sheets may disappear or may never even appear, and hence seem to be missing. In such cases individual muscles derived from the sheet seem to be very distinctly separate from one another. Indeed, in human anatomy they are often given separate names and this hides the true commonality of their original developmental and evolutionary relationships.

Let us examine each of the myotomal components in order to see this in detail.

The Dorsal Myotomal Block, the Epaxial Muscles

The dorsal part of the myotome, influenced by molecular factors emanating from the neural tube, becomes, at each segmental level, the precursors of those parts of the voluntary muscles of the trunk that are initially lying dorsal to the developing nervous system and deep to the dermatome. It first forms a sheet of muscle called the epaxial (dorsal to the axis – notochord) muscle block that eventually divides into the complex spinal extensor musculature of the adult (though this musculature performs many more functions than just spinal extension). As described above, it comes to form deep and superficial muscle blocks. Thus the eventual resulting spinal extensor muscles that are produced are recognizable as two primary layers: deep and superficial.

The deep sheet of muscle lies, on each side, within the cranio-caudal gutter formed by axial spines and lateral transverse processes that are bony projections developing on each vertebra (see the sclerotome, below). This deep sheet gradually differentiates into three layers of individual muscles. The very deepest layer usually represents only a single somite, the eventual muscle thus passing from one vertebral segment to the next. The muscles developing from the middle, more superficial layer (but still very deep in the trunk) cross more segments (say two to three). The muscles of the most superficial layer of all (though again still very deep, and also lying, as they do, in the vertebral gutter), may cross as many as five or more segments. Overall the muscles of these layers are called the transversospinalis muscle (with separate names for each of the above components). Most of the muscle fibers are orientated obliquely from lateral (e.g. vertebral transverse processes) above, to medial (e.g. spinous processes) below.

External to these deep muscles is the more muscular superficial sheet mentioned above. This sheet lies in a much wider, shallower, gutter that extends from the vertebral spines (medially) to the angles of the ribs (laterally) in the thorax (or equivalent bony elements in other trunk regions – for more about these bony parts, see sclerotome, below). Again, most of these muscle fibers are oriented obliquely from lateral above to medial below.

This muscular overlay also differentiates into three portions, but these are arranged medially, intermediately and laterally on the back. The individual fiber bundles of these muscles cross greater numbers of vertebral segments than the deep muscles. They are called, overall, the erector spinae muscles with, again, separate names for each of the three components just described. In extending from cranial to caudal in the trunk they pass from the occipital bone of the skull, into the neck and along most of the vertebral column. Though with many separate names in human anatomy they are not, basically, different muscles, but merely developmentally a part of this same pattern.

Both of these dorsal muscle blocks are shown longitudinally in Fig. 4.3 (a,b) and cross-sectionally in Fig. 4.4.

The muscle bundles of both deep and superficial sheets act together in extending the spine; hence the overall title: spinal extensor muscles. However, the oblique directions of most of the muscle bundles indicate that, when the bundles of one side are acting alone, or alternately with the bundles of the other, they aid lateral bending and axial twisting of the spine. Thus, though extension is clearly important, it is likely that the bending and twisting functions are functionally more important in humans, and often less usually recognized.

Such an arrangement, in addition to bending and twisting the vertebral column, also produces relief of compressive strain on the bony column. This is a 'tensegrity' mechanism (described elsewhere) and this, too, may be functionally very important. This is not to say that this mechanism is the sole component of compression bearing in the spinal column; a large compression mechanism is truly that of the vertebral bodies, intervertebral joints, and the intervertebral disks. However, the compression that these components bear is reduced by the tensegrity functions of the vertebral muscles and their tendons, ligaments and fascial sheets. This function is not a function of simple longitudinal muscles, but due to the oblique nature of their arrangements. Indeed, it is this component of compression bearing that is frequently

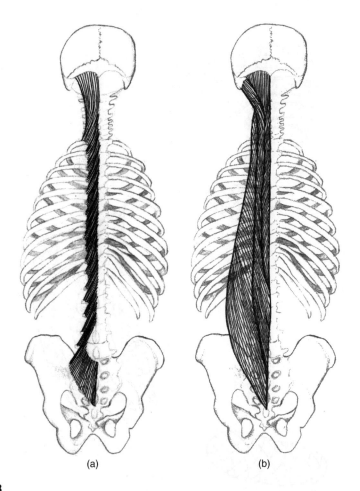

(a)   (b)

FIGURE 4.3

Longitudinal view of the muscles that are eventually derived from the deepest portion (a) and more superficial portion (b) of the hypaxial myotome in the trunk.

interfered with through damage to these muscles and their associated tendons, bony attachments, surrounding fascial sheets, and internal myo-fascial components.

Although these dorsal trunk muscles are all embryologically truly dorsal derivatives, they eventually come to be covered over by other muscles that are really ventral in origin (from the hypaxial myotome, see below), but that migrate dorsally at later stages in development. That is, notwithstanding the eventual very dorsal position of these latter ventral muscles (and the fact that human anatomists have generally given them names that imply that they are dorsal, for example: latissimus dorsi) they are, in fact, truly ventral derivatives in relation to limbs (which themselves are ventral trunk derivatives). Thus, these limb muscles, as we shall see later,

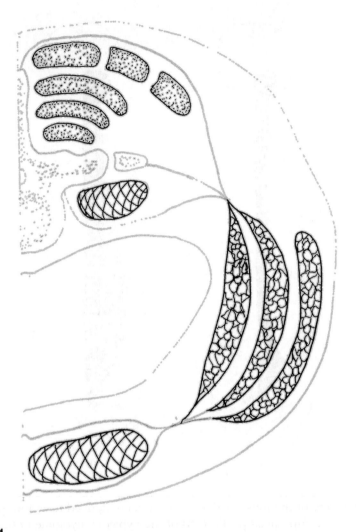

FIGURE 4.4

Cross-section showing the muscles as represented in the abdomen: the various dorsal epaxial muscles on the dorsal aspect of the vertebra, and the embryologically ventral hypaxial muscles: quadratus lumborum dorsally, rectus abdominis ventrally, and the three abdominal oblique muscles laterally. This diagram is 'exploded' so that the various components can be more easily seen.

are derived from the ventral (hypaxial) myotome of the trunk. Examples include, from the upper limb, latissimus dorsi, and from the lower limb, gluteus maximus.

Just to further complicate matters, the muscles of the limbs, derived only from ventral trunk precursors, are themselves split into dorsal and ventral components within the limbs. Understanding all this is necessary to explain the pattern of muscle attachments, the complexities of their nerve supplies, the reflex actions and movements in which they are involved, and so on. We will go into the limb muscle complexities in a later section.

The Ventral Myotomal Block, the Hypaxial Muscles

The more ventral parts of the myotome of each segment of the trunk lying next to the more ventral axial notochord are influenced by molecular factors secreted from the notochord to become the ventral body wall muscles. This is a much bigger block of pre-muscular tissue called the hypaxial (below – ventral – to the axis: notochord) myotome. It becomes a complex of muscle sheets forming the ventral portion of the muscular body wall of the trunk: the neck, thorax and abdomen. However, it is so large that it also extends far laterally to form the lateral body wall. And, as we have seen, some components actually extend so far dorsally that they come to form part of the dorsal wall of the abdomen.

Like the epaxial muscles, the hypaxial muscle block also differentiates into longitudinally arranged columns, and relatively horizontally arranged deep to superficial layers. The longitudinal columns of muscle form at the back and front of the body cavity. The more horizontally arranged layers form the major portion of the dorsal, lateral and ventral trunk walls of the neck, thorax and abdomen (Figs. 4.4 and 4.5).

Again, the nature of the patterning of the body is such that, like the epaxial muscles, their blueprint progenitors extend throughout the trunk (even though, in particular regions, they appear to be absent, but see below). Once again, then, understanding the arrangements in any one segment, provides information relevant to most other segments.

The Dorsal Longitudinal Hypaxial Muscles

This muscular group lies in a dorsal position but lateral to the developing vertebral column. In many vertebrates, this is a complete column passing from the neck to the pelvis (the prevertebral muscle). In humans, it is obvious in the abdomen as the quadratus lumborum muscle (and also as the psoas minor muscle). Quadratus lumborum becomes a series of vertical muscle bundles crossing several segments from the last rib to the pelvis. Its chief function is in supporting the back of the abdomen. The psoas minor muscle has become commandeered, as it were, by the lower limb and its main action relates to hip movements. However, it is actually a dorsal derivative of the trunk hypaxial muscle block.

In the neck, the dorsal component of the hypaxial muscle block is evident as the longus cervicis and longus capitis muscles.

In the thorax this dorsal muscular column is generally absent. It does not form where the extended thoracic rib system subdivides the space it would occupy. The overall pattern is shown, however, by the fact that there are, in some individuals, small vertical muscular slips present as variations on the internal surface of the back of the thoracic wall. It is likely that these are the remnants in the human thorax of the complete muscle sheet that is found in many parts of the trunk in so many

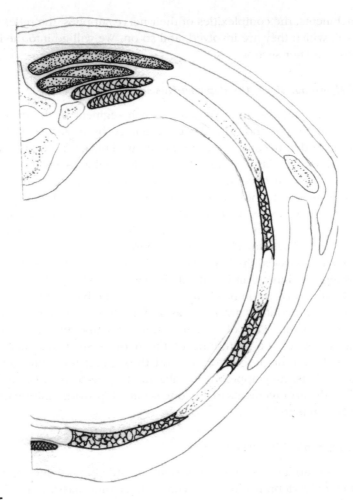

FIGURE 4.5

Cross-section showing the muscles of Fig. 4.44 as represented in the thorax. There are a few small muscle fibers on the dorsum of the thorax that are the equivalent of the very large quadratus lumborum in the abdomen. There may also sometimes be very small fibers on the ventral aspect of the thorax on the sternum that are the equivalent of the very large rectus abdominis in the abdomen. The lateral oblique muscles are subdivided by the ribs. They are in three layers between the ribs (but the layers are not shown in the diagram).

other animals. It is in this way that the comparative pictures, and especially human variations, come to be understood by developmental patterning and comparative variation.

The Ventral Longitudinal Hypaxial Muscles

The most ventral column of the hypaxial muscle group is also a longitudinally oriented group of muscle fibers. Again, these represent many segments, and lie close to the midline ventrally (Fig. 4.6). Again, too, the pattern varies with the region of the body.

In the abdomen it becomes a powerful longitudinally orientated set of fibers, the rectus abdominis muscle, with intermediate tendons (said by some to represent remnants of the original segmental separations – but I doubt this is really so – tendons are so important functionally and are produced by different molecular processes and maintained by different mechanical factors).

In the neck, the serial homologs of this longitudinal block differentiate into the various longitudinal ventral neck muscles. These include the (ventrally located) intrinsic muscles of the tongue (occipital somites, discussed with the head), and muscles in the neck above the hyoid bone (also discussed with the head). More caudally in the neck proper, these include the ventral muscles from the hyoid bone to the sternum (cervical segmental infrahyoid muscles such as the sternohyoid and sternothyrohyoid muscles).

As with the dorsal muscular column, the thoracic component of the ventral column is usually absent where the sternum lies in the position it would occupy. Also like the dorsal column, however, the basic overall pattern is indicated by occasional variations. Thus some vertically oriented muscle bundles may be present lying on the sternum. When present they are called the rectus thoracis (following the terminology of the rectus abdominis in the abdomen). In other words, there is, again, cranio-caudal continuity of structure here. It is just that, in humans, the thoracic portions are usually suppressed. Summaries of these muscles are presented in Figs. 4.4, 4.5, and 4.6).

In this regard, it must be mentioned that there is another set of vertical paired muscles that may be present on the sternum as a variation in humans. They are not, however, a persistent rectus thoracis, but rather persistent portions of another much more superficial muscle of dermal origin, the panniculus carnosus. This is a large muscle sheet in many mammals, found in the superficial fascia, and not therefore of dermatomal origin). It is particularly present in monkeys and apes, but is largely suppressed in humans (Figs. 4.7(a,b) and 4.8(a–c).

The Lateral (Oblique) Hypaxial Muscles

The lateral, more horizontally arranged components of the ventral myotomal block in the trunk differentiate into three layers arranged as superficial, intermediate and deep. This is so in all three portions of the trunk: neck, thorax and abdomen. In both the abdomen and thorax, they are named the external, internal and innermost muscular layers.

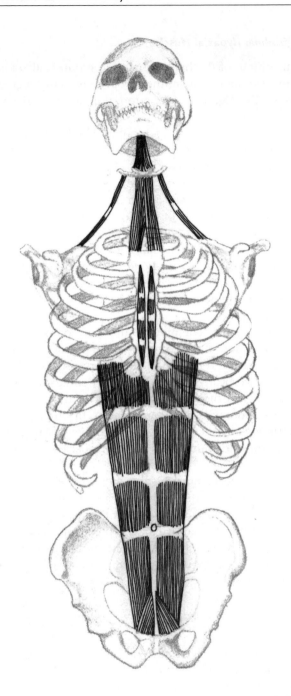

FIGURE 4.6

Longitudinal view of the ventral components of the hypaxial myotome along the trunk: in the head and neck, the various hyoid muscles, in the thorax the rectus thoracis (when present), and in the abdomen the very large rectus abdominis and its caudal component, pyramidalis.

(a)   (b)

FIGURE 4.7

A few muscle fibers may sometimes be seen on the ventral surface of the sternum that are not part of the myotomal rectus thoracis, but that are part of the muscle found in the superficial fascia. Another such variant can also be found in the axilla. These are not myotomal derivatives, but rarely found remnants of what is a large muscle sheet, the panniculus carnosus, in the superficial fascia in many other animals. Some of these are constantly found: the dartos muscle in the scrotum and a few muscle fibers in the labia majora (the equivalent in the female of the scrotum in the male).

In the abdomen on each side they form complete sheets, and are specifically named the external and internal oblique muscles, and the transversus abdominis muscle (Figs. 4.2 and 4.4 and 4.5).

In the thorax, they are divided by the presence of intervening ribs in each segment. As a result, they are named the external, internal and innermost intercostal muscles lying in the gaps between the ribs. Though a completely separate muscle at each inter-costal level, they are, as a group, the serial homologs of the complete oblique muscle sheets evident in the abdomen.

In the neck they are evident as the three scalene muscles (called anterior, middle and posterior in human terminology). It would probably be taking the serial homologs too far to suppose that these three muscles in the neck represent the three layers in the abdomen and pelvis.

Though in this discussion of these muscles, it appears that ribs are present only in the thorax, they are actually present at every segmental level. It is only in the thorax that they exist as separate recognizable bones curving around a large part of the thorax. There is the caveat that, as developmental variations, there may be one, two or even three separate free-standing ribs in the neck (cervical ribs), and one or even two such ribs (lumbar ribs) in the abdomen. In these other regions, the hindmost part of the head (occipital region, see later), and the neck, abdomen and pelvis, rib elements are truly present as part of the overall repetitive pattern. But in these regions they are generally not separate bones but developmentally present as: the lateral components of part of the occipital bones of the skull, the shorter ventral portions of the transverse processes of the cervical and lumbar vertebrae, and the lateral masses of the several segments that eventually comprise the adult sacrum. This is further explained below with the treatment of the sclerotome and the developing skeleton.

Though, again, any one set of these ventral trunk muscles has functions that can be worked out in relation to their precise origins and insertions, and tested through experimental electromyography, their major overall actions relate to the trunk as a whole. Thus they function in external movements of the trunk, such as in locomotion where they participate in flexion of the trunk when acting at the same time, and bending and twisting of the trunk when acting alternately.

Yet they have many other important functions. They act during internal movements of the trunk, for example, changing the internal pressure in the body cavities. This occurs especially during acts such as forced respiration, forced micturition, forced defecation and powerful parturition, and even, to lesser degrees, in such smaller expulsive movements as coughing, sneezing and vomiting. These body cavity pressure effects can also be called into play during some voluntary actions of the external trunk, such as helping stabilize the trunk during heavy weight-lifting, and so on. In weight-lifting, their actions may be enhanced using strong external belts to reinforce them. The tension in such belts reduces the tension needed in the muscles and other bodily tissues (another example of tensegrity) to produce compression.

Summary of Myotomal Derivatives

In summary, then, though we can identify, in the adult trunk, a very large number of muscles with a multitude of names in what might seem to be confusing complexity, this complexity is reduced by understanding the developmental processes by which they have arisen, and the serial homologs that they represent. The simplicity of developmental origins is evident through the processes of separating (and/or fusing) of structures in relation to longitudinal segmentation, dorsal and ventral subdivision, and superficial and deep separation, and also in the reverse: suppression or loss of elements.

These various developmental processes also give rise, in other species, not only to muscles generally present in adult humans, but also to muscles that are usually absent in humans. Their presence 'on occasion' in humans is evidence of the commonality of the overall pattern, and the way in which it may be modulated in different species. Such human variations and their comparison with other species (see later) help not only in understanding human anatomy but also in providing new hypotheses for developmental biologists.

The Nerve Supply of the Myotomes

Though we might expect the nerves (and also arteries, veins and lymphatics) of these developing muscles to be covered in separate sections, in fact, they are best understood with reference to the preceding picture. Thus each segmental spinal nerve, after exiting the vertebral canal, undergoes a primary division into what are called dorsal and ventral primary rami. In other words, each shares the same dorso-ventral division as the myotome. It all goes together so well.

The sequential dorsal nerve rami supply all the muscles arising from the dorsal portion of the myotomes (epaxial myotomes), the ventral nerve rami those arising from the ventral portion of the myotomes (hypaxial myotomes). The few muscles that arise from a single segment (somite) receive a twig from the appropriate ramus of that segment. But because the great majority of muscles in the trunk develop, as we have seen above, through the coalescence of several segments, they, appropriately, receive nerve supplies from each of the several corresponding spinal nerves. In the trunk, these are usually separate nerve branches at each level, for example, the several lumbar spinal nerves, each giving branches sequentially to the various parts of the oblique and rectus muscles of the abdomen. Sometimes, however, such branches join together to form a single nerve before supplying a muscle that covers (represents) several segments. A good example here is the long nerve from the union of muscular nerves of several thoracic segments that supply the multisegmental serratus anterior muscle (see later).

The fact that some individual muscles result from several segments and therefore have a nerve supply that includes several somites, means, of course, that the movements that such muscles produce are brought about by the activity of a group of segmental spinal nerves. The pattern of nerve segments that supplies a particular muscle is thus also responsible for producing the equivalent pattern of movement resulting from the activity of that muscle. The presence of these patterns enormously relieves us of feats of memory. Thus function (movements produced) can be largely worked out if we understand the underlying structural patterns. This is, however, most clear, and also most useful, in understanding limb structure and function (see below).

Vascular Arrangements to Myotomes

Similar arrangements (though usually somewhat less clear) can be discerned in arterial, venous and lymphatic vessel organizations. Thus, in the trunk, each segment receives its own arterial supply (intercostal arteries for each segment in the thorax, lumbar arteries for each segment in the abdomen). Likewise each segment takes up its own venous and lymphatic drainage (equivalently named veins and lymphatics). Even the lymphatic vessels have a somewhat similar set of arrangements.

However, in all of these systems there are marked longitudinal anastomoses up and down the segmental levels, and in and out between the various superficial and deep layers (see below). Because these systems arise from different molecular mechanisms than muscles there are further complexities of pattern to be considered. They will be covered in more detail below.

The dorso-ventral division of the myotomal part of the somite in the trunk just described is fundamental and obvious. This mechanism applies equally, however, to the other somite derivatives, the sclerotome and the dermatome. Thus, they too, come, in the trunk, to have dorsal and ventral components. Let us look at the sclerotome next.

4.1.3 The Fate of the Sclerotomes

The second primary element of the somite is the sclerotome. Because the sclerotome gives rise, among other things, to bone, and bones are very easily examined, the greatest amount of developmental information exists for the skeleton. Being largely intersegmental structures, the sclerotomal patterns are somewhat different from the myotomal.

Thus, the developing sclerotomes, and the bones and joints that they eventually become, though they split like the muscles into dorsal and ventral moieties, differ in that they are not segmental but intersegmental in position. That is, they lie between the segmental muscles. This makes functional sense when we realize that the segmental muscles must pass between the bones (vertebrae, ribs), one above and one below, that they move.

Thus on each side, the caudal part of the dorsal sclerotome of one segment unites with the cranial part of the dorsal sclerotome of the segment below. The two together, as a compound structure, become the dorsal part of each pre-vertebra. These developments are influenced by factors from the developing nervous system so that the bony elements that develop from them, the vertebral spines, laminae, pedicles, articular processes and dorsal parts of the transverse processes, form a segmented but articulated skeletal arch lateral and dorsal to the spinal cord.

Similarly, the ventral parts of two adjacent segments unite to form the remaining parts of the pre-vertebra. These include most of the bone of the vertebral body, the costal (costal equals side equals rib) processes, the ventral portions of the transverse processes (also costal in origin but not separate), the true free ribs in those regions where ribs are present (largely the thorax), and segmented ventral elements on each side of the ventral midline of the trunk (putative sternum).

These last elements in this series, special paired ventral components, soon fuse across the midline so that a series of central sternal segments can be discerned, the sternebrae. And even later during growth these individual sternebrae become fused into a single bony structure, the sternum. In many creatures these sternebrae remain separate throughout life, even in close evolutionary relatives such as chimpanzees and gorillas. Further, this fusion sometimes does not occur in humans if there is great delay in growth (for whatever reason). Thus such individuals may remain with partially fused or even unfused sternebrae, and thus may spuriously resemble, in this feature, apes and monkeys.

In summary, however, like the vertebral canal dorsally, these various components come to form a segmented ventral skeletal tube around the viscera in the thorax. This elaboration of the costal system is not obvious in the human head, neck, abdomen and pelvis where the costal system is much abbreviated. It is present only as the ventral parts of the lateral masses of the occipital bone (in the head), the ventral parts of the transverse processes in the neck and abdomen, and the ventral part of the lateral masses of the sacrum (in the pelvis). Nevertheless, though apparently different in these regions, the underlying pattern is actually common to all segmental levels. Indeed, there are sometimes ribs in the lower few cervical and upper few lumbar vertebrae (and in some conditions of growth deficiencies, non-fusion of lateral masses of the sacrum in the pelvis).

These ventral developments of the vertebral bodies occur under the influence of molecular factors from the neighbouring notochord. The skeletal axis (the notochord) is eventually replaced by them. The notochord disappears, though some workers aver that the nucleus pulposus in the center of the intervertebral disks is a remnant of it.

In addition to bone, many other elements are related to the sclerotome. Thus junctions between individual elements of the sclerotome may become joints of variable structure. Connections of all parts of the sclerotome to one another may be evident as fascial sheets. When a fascial sheet comes into contact with bone it is called a periosteum, with cartilage, a perichondrium, and with muscles, a perimysium. Although possessing these separate names they actually form a complete single closed system. As a result, in between the structures that they cover, they are called fascial sheets. Further, they are specifically called deep fascial sheets. This is because two other sets of fascia can be defined. There is a set of connective tissue sheets, the superficial fascia, that develops more peripherally under the skin (see later), and a third set of sheets, the visceral fascia, that develops more centrally, lining the body cavities and surrounding the internal organs (again, see later).

These connective tissue elements of the sclerotome, the various fascial sheets, also come to cover and support the various nerves (and arteries, veins and lymphatics) as they develop.

The Nerves of the Sclerotomes (and Therefore of Connective Tissues)

The same paired nerves that supply myotomes also supply the eventual derivatives of the sclerotomes. Thus, as before, each segment receives its own nerve branches from the dorsal and ventral rami of the spinal nerves. It likewise receives its own components from the sympathetic chain. The motor fibers are mainly those to the smooth muscle of the blood vessels lying along the fascial sheets. The sensory fibers are mostly those to the muscles, tendons, ligaments and fascial sheets, and are mainly involved in proprioception and pain. Indeed, there are a surprising number of proprioceptive and pain receptors in the various periosteal, fascial sheets, ligaments and the tendons of muscles around the spinal column. This may account in part for the importance of posture and movement in the spinal column, and the frequency of spinal strain, cramp and pain when these tension-bearing elements are damaged.

Likewise, each vertebral segment in the trunk has its own paired arteries (intercostal arteries for each segment in the thorax), lumbar arteries (for each segment in the abdomen), and sacral arteries (for each segment of the sacrum). The neck differs in that the main vessels to and from the head traverse the neck so that direct branches from main vessels generally supply closely located neck structures.

Likewise each segment takes up its own venous and lymphatic drainage (equivalently named veins and lymphatics).

In all of these systems, however, there are marked anastomoses between segmental levels, and between superficial and deep systems. Such anastomoses especially involve arteries, veins and lymphatics around each vertebra externally, and within each vertebra both within the vertebral canal containing the spinal cord, and within the body of each vertebra in the marrow cavities. There are major anastomoses between these three sets of vessels in each segment. There are also large longitudinal anastomoses between spinal levels up and down the vertebral column. These various anastomoses are so complex and form such large lakes of blood, that they are especially the sites of neoplastic metastases.

Again, because these systems arise from different molecular mechanisms than muscles, there are further complexities of pattern to be considered. These will be covered in more detail below.

4.1.4 The Fate of the Dermatomes

The dermatome of each trunk segment becomes eventually that part of the skin (dermis) that underlies the epithelium derived from the external ectoderm. As with the myotome, the dorsal part of each dermatome is small. It gives rise to the dorsal dermis of the trunk, but only that smaller portion near the midline.

As was the case for the muscles, the ventral part of the dermatome is much larger. It is, likewise, associated not only with the ventral trunk skin, but also the lateral skin and even most of the dorsal skin.

None of the dermatomes are as clearly identifiable structurally at each segment level as are the muscles. It is just that they are organized as a column along the length of the trunk. The areas representing individual segments are thus arranged as a series of girdles around the trunk. They have no clear boundaries, overlapping one another to a large degree. The evidence for dermatomal patterns lies mainly in understanding their (cutaneous) nerve supply.

Because, in humans, the dermatome is largely related to the skin and does not have visible boundaries, it tends to be forgotten that it is actually a structure in its own right. It thus stems from the peripheral mesoderm; it is therefore superficial in position, superficial, that is, to the derivatives of the myotome and the sclerotome. This superficiality is not just a matter of anatomical description; it an important developmental difference. Thus the myotomal and sclerotomal derivatives are all bounded on their exterior by the outermost layer of the deep fascia. In contrast, the dermatomal structures lie within fascia that is external to the deep fascia. That is: they lie within a connective tissue layer called the superficial fascia. Secondly, though in humans we generally think of the skin as just dermis and epidermis, it actually also includes a similar series of structures (such as muscles, bones, nerves, vessels) as the deeper parts of the trunk.

The main components of the dermis are the roots of the epidermis, and the supporting connective tissues. However, the dermis also contains the bodies of those epidermal glands that have invaginated from the epithelium itself to become larger structures (much in the way that the pancreas is an invaginated gland from the epithelium of the gut – see below). Such structures include the mammary glands in the trunk and the salivary glands in the head.

Dermal Bone

Further, however, just as bone forms within the deeper mesoderm, so too, bone can be found in superficial mesoderm. In many animals there are large bony structures in the superficial mesoderm. These bones are called 'membrane bones' because they develop directly in the mesenchymal membrane of the peripheral mesoderm without going through a stage involving a cartilaginous model. In humans, these components are very small in the trunk (though large in the head – see later). Thus the lateral part of the clavicle arises 'in membrane' without going through a cartilaginous precursor (the remaining parts of the clavicle arising in relation to the sclerotome of the upper limb girdle). In some mammals (e.g. kangaroos) and reptiles, an equivalent bony element of the lower limb girdle, the supra-pubic bone, arises in the same way.

The special superficial nature of these bones is again evident as variations in humans. Thus some of the superficial cutaneous nerves of the neck and upper chest (cervical nerves) that lie external to the deep fascia may actually pierce the superficially derived portion of the clavicle.

This information would seem fairly inconsequential. It is, however, part of the overall developmental pattern and helps in anatomical understanding. Thus a precisely similar pattern is very much present in the head. However, as we shall see later, the pattern in the head is reversed in that a very large portion of the head skeleton arises as a superficial 'membrane' bone superstructure (the cranial vault) latched onto the deeper endochondrally derived bone of the sclerotome (the cranial base).

Dermal Muscle

Yet another derivative of the peripheral mesoderm that is often ignored is voluntary muscle. Thus, although most of the musculature of the trunk is derived from the myotomes, some muscles originate *in situ* in the peripheral mesoderm, in superficial fascia. They can be thought of as the muscular equivalent of dermal bone.

Again these muscular elements in the trunk are rather small and would seem to be inconsequential. However, in many mammals, these muscles, going by the name of the panniculus carnosus muscular sheet, are very extensive (Fig. 4.7) and have important functions. Thus, the flicking skin of the cow drives away flies; the curvature induced in the skin of the bat wing provides additional aerodynamic factors in flying; some equivalent muscles in kangaroos form the wall of the pouch; and in the skin of swimming mammals these muscles may add to their improved skin aquadynamics, and so on. In humans they are mostly small occasional variations that can be found in relation to the axilla, the groin, the ventral thoracic wall, and the superficial genitalia (Fig. 4.8).

Though apparently inconsequential in the human trunk, they become enormously more strongly developed in the human head, giving rise to the muscles of facial expression that also lie in the superficial fascia external to such structures as lie under the deep fascia in the head (e.g. the muscles of mastication: see later). Again, therefore this helps to indicate the continuity of developmental patterns, in the head as in the trunk.

As we come to understand the superficial components of the dermis, we then realize that this pattern applies to all the other structures. In addition to the main arteries, there are minor arteries that lie in the superficial fascia. Knowledge of these is critical in clinical anatomy, for example, in knowing what portions of the skin are nourished by specific arteries – important in understanding the large skin grafts that are sometimes required following major trauma.

Likewise, the veins of the superficial fascia form a superficial network that is largely, though not completely, separate from the deep veins that accompany the major arteries inside the deep fascia. These superficial veins do have occasional connections with the deep venous system and knowledge of these connections becomes important in relation to such matters as venous thromboses and varicose veins.

The lymphatic vessels and their nodes are likewise arranged as superficial and deep sets, and recognition of this is important in many clinical situations.

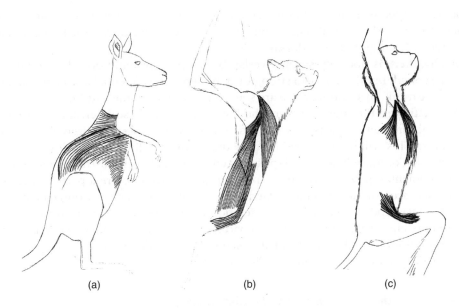

(a)   (b)   (c)

FIGURE 4.8

The more extensive dermal muscles of the superficial fascia in the trunk of some other species: (a) kangaroo, (b) bat and (c) rhesus monkey.

In addition to these various clinical importances, however, all of this indicates the enormous consistency of human anatomy and human development.

The Nerves of the Dermatomes (and Therefore of Skin)

The nerves supplying the dermatomes forming the skin are also organised as a series of cutaneous branches from the main branches of the spinal nerves from each segment. Thus there are small mixed (which means containing both sensory and motor elements) dorsal branches from the dorsal primary ramus of each spinal nerve, and these, after giving their muscular branches to the true dorsal muscles (the epaxial muscles) provide cutaneous branches supplying the skin near the dorsal midline. Of course, the branches to the muscles, muscular branches, also contain sensory as well as motor fibers – there are sense organs in relation to muscles and their tendons. Likewise the cutaneous branches also contain both sensory and motor fibers, obviously sensory fibers in relation to various aspects of skin sensation, but also motor and secreto-motor fibers to some special striated muscles in the superficial fascia, to smooth muscles (motor for example in blood vessels and in relation to the muscles that erect the hairs, and secreto-motor to such structures as sweat glands in the dermis).

In like manner, but much more extensively, the ventral primary rami have lateral and ventral branches that provide muscular branches to the hypaxial ventral muscles as already described and then, more peripherally and subsequently, smaller cutaneous branches to the ventral skin. Because, however, the ventral skin not only covers the ventral regions but also extends far dorsally, these branches are considerably larger than the dorsal.

Finally, there is not only strong overlap between ventral and dorsal components just described, but there is also strong overlap between sequential segments up and down the trunk. In addition, there are some interruptions in the pattern of sequential segmentation in the regions where the limbs develop. These interruptions appear as though the ventral parts of some individual segments were entirely drawn into the limbs. This will be described in more detail with the limbs (see below).

As a result of these heavy overlaps between segmental nerve distributions, at least three segmental nerves would have to be cut or damaged before a particular single segmental area of skin would be denervated. These segmental arrangements are nevertheless rendered clear by consideration of such sensory skin disturbances in conditions like herpes zoster, where a virus lodged in a particular segment of the spinal cord induces the nerves of that segment to produce a pattern of pain and blisters that lie in that particular girdle around the trunk.

The Vessels of the Dermatome (and Therefore of Skin)

The arteries, veins and lymphatics are also arranged as superficial branches of the relevant spinal vessels that supply the approximate areas of the dermatomes. Again, the segmental arrangement is clearest for arteries, somewhat less clear for veins because of the increased anastomoses between superficial and deep veins, and least obvious of all for the lymphatics because these are most heavily influenced by the much greater pattern of lymphatics related to the development of the limbs (see later). Similar longitudinal and deep/superficial anastomoses as before are also evident.

In this way we now have all the primary elements of the body wall and can see how this provides understanding of the adult anatomical arrangements.

4.2 The Internal Trunk: From Shell to Framework

In the prior part of this discussion, the external trunk was largely described as though it was a modification of a more or less similar set of segments along the length of the early fetus. The internal trunk differs in that, although it is more or less similar along the length of the fetus, it does not display a segmented structure. However, during the further growth and development of the fetus, changes in the internal trunk occur that mean that the various parts become separately differentiated, gradually changing towards what we find in the adult.

4.2.1 The Two Basic Internal Trunk Building Blocks: Mesoderm and Endoderm

We already know that the basic structures of the internal trunk with its internal organs develop from the endoderm and the deep parts of the lateral plate mesoderm (see Fig. 4.2 b). The endoderm gives rise to the epithelial layers of the gut tube and its glands. The deep parts of the lateral plate mesoderm produce the supporting structures (connective tissues and smooth muscles) of the gut tube and its glands, as well as other structures we will examine later.

Developmental factors modulate the simple tubular gut into the eventual complex alimentary canal. These changes include: marked elongation so that the tube becomes complexly coiled, and marked architectural differentiation so that the tube has a variety of different functional and structural sections. Thus a complex gut develops, with biochemical maturation of the secretory and absorptive epithelia, the formation of the small glands in the epithelium, and the extrusion of much larger glands both into the thicker gut wall and completely outside the gut wall (see later).

All glands maintain, however, their epithelial connection with the interior of the gut tube through the ducts that carry products from the glands into the gut. The lateral plate mesoderm provides the various connective tissue and muscle coats that support the epithelial gut tube and form a backing for the serous membranes that cover the various structures, separating them from the serous cavities.

Developmental factors also modulate the cranial end of the gut tube so that respiratory pathways (gas-carrying structures, eventually the trachea, bronchi and bronchioles of the adult lung) and respiratory sacs (gas-exchanging structures, eventually the alveolar ducts and alveolar sacs of the adult lung) develop. Even though these structures eventually lie in the thorax, their initial development is a matter for the head which we will consider later. However, as these tubes later expand and develop, they come to form the air passages and gas-exchange surfaces of the lungs, and in particular they come to grow caudally into the developing thorax. Accordingly then, at this stage they become important in understanding trunk arrangements.

4.2.2 The Fate of the Gut Tube

Though there is no cranio-caudal segmentation of the internal trunk, as there is in the external trunk, there is a simple longitudinal differentiation of the gut tube. Three nominal gut parts become identifiable from the endoderm alone: foregut, midgut and hindgut. In addition, two other parts form in relation to the two parts of the endodermal gut tube that abut at each end against the ectoderm without intervening mesoderm. These are, at the cranial end, the oral plate (oropharyngeal membrane, see section on the head), and at the caudal end, the cloacal plate (cloacal membrane, see later this section). These parts are not just named for descriptive convenience but reflect underlying developmental arrangements.

Thus, this differentiation of what is originally a simple gut tube is driven by the orderly expression of groups of Hox genes. Development of the gut tube proper involves continuous elongation. This becomes so great that the gut becomes heavily coiled and, at one stage, is so large for the small body cavity of the fetus, that it herniates out past the (at that stage) incomplete external body wall. It later returns to the internal body cavity and undergoes rotation and folding, thus becoming neatly packed into the body cavity.

This endodermal tube continuously interacts with the mesenchyme of the deep part of the lateral plate mesoderm that surrounds it. These interactions help determine the specific characters of the gut endoderm in the various parts of the tube (e.g. foregut: esophagus, stomach, cranial part of the duodenum; midgut: caudal part of the duodenum, jejunum, ileum, and parts of the colon; hindgut: rest of colon and rectum). Thus, the endoderm produces: simple small glands directly in the gut epithelium, larger and more complex glands in the supporting thickened connective tissues of the mesenchymal wall of the gut, and extension of some of these glands beyond the gut wall to form major digestive organs (e.g. a component of the pancreas, a component of the liver, and so on). These last become large structures apparently quite separate from the gut tube. But the evidence of their original gut derivation is still present as their ducts still connect to the gut and provide pathways for digestive secretions to pass into the gut tube.

These complexities are under the control of molecular factors (such as sonic hedgehog) working through other factors (e.g. bone morphogenetic protein 4) in the adjacent deep lateral plate mesoderm. This deep lateral plate mesoderm also provides the connective tissue frameworks and carries the nerves, blood vessels and lymphatics to all the endodermal structures.

4.2.3 Up- and Down-Regulation of the Gut

Thus the alimentary canal is a highly convoluted pathway carrying and breaking down foodstuffs to support the digestive functions of the organism. Along with this, however, are mechanisms to handle various pathogenic organisms (especially bacterial) that are taken in with the food, so that it also has an immune function. The alimentary tract is, further, a dynamic organ that responds dramatically to changes. As a result it can produce a wide variety of movements (e.g. churning, peristalsis, mass peristalsis, and reverse peristalsis) as well as handling chemical activity by changes in secretions and absorption of foods and fluids.

Further, the gut mucosa can respond to mechanical and digestive demands by changed (heightened) metabolic activity – a mechanism known as up-regulation. Normally the mucosal cells, differently in the different regions of course, are shed into the lumen so that the entire intestinal surface may be turned over in a few weeks. During up-regulation this process may be greatly increased to meet immediate physiological demands. Up-regulation is fully reversible. Thus, after a meal, the mucosa returns to a resting state – an overall response called down-regulation.

These up- and down-regulatory changes may well occur at different times in different portions of the gut tube depending upon the stage of digestion, and the timing effects of different types of foods.

These responses may sometimes be long lasting and difficult to reverse. Thus in under-nutrition, and especially in severe malnutrition, down-regulation may be so profound that it may be dangerous to treat such individuals by the natural immediate response of giving lots of good food. Reversing malnutrition requires to be tackled slowly and carefully so that up-regulation, which has been profoundly depressed, has time and ability to gradually reverse.

There are other structures that develop in relation to the gut tube at the cranial and caudal ends where the gut tube opens to the exterior through the facial region in the head and through the perineal region in the trunk. Thus the eventual external openings of the gut tube are the mouth and anus. These openings are located alongside the openings of other tubes, the respiratory system in the head, and the urino-genital system in the perineum. They are discussed later.

4.2.4 The Body Cavity

In the preceding, mention has been made, several times, of the body cavity. This is a cavity that, in general terms, separates the external trunk from the internal trunk. We already know that the lateral plate mesoderm consists of a superficial portion that lines the internal surface of the external body wall. We also know that the deep portion of the lateral plate mesoderm covers and supports the endodermal structures of the internal trunk). We have also already described how the separation of these two parts of the lateral plate mesoderm in the trunk produces between them a cavity (Fig. 4.2(b)). This cavity later becomes divided, by processes we will examine later, into separate pleural, pericardial, peritoneal, and other smaller spaces of the adult.

The external lining of the serous cavity is the superficial layer of lateral plate mesoderm. It is called the parietal (parietes: wall) layer consisting of a backing of connective tissue and a lining of a serous mesothelium. Though it is part of the internal trunk, it is intimately adherent to the inside of the external trunk. It receives its vascular and nerve supplies from the paired components (paired spinal nerves, paired trunk arteries, etc) of the external trunk passing into it from the external trunk wall. This is why, for instance, the parietal serous membranes (later peritoneum, pericardium and pleura in relation to the corresponding cavities) are seriously and sharply painful when inflamed.

The internal lining of the serous cavity is the deep layer of the lateral plate mesoderm. It thus covers the developing gut (hence: visceral layer of the serous membrane). Although these layers are given these separate names, they are in fact a single sheet of mesothelium that becomes folded upon itself as various organs become invaginated into the cavity (in the way that one can invaginate one's fist into a partially inflated balloon). The nerves and vessels of the visceral serous layers come, of course, from the midline nerves and vessels that supply the developing gut. Pain originating in the visceral serous membrane, from, say, inflammation, is therefore much less sharp and much less easily localized than from the parietal serous membrane (as described above). The former inflammation is sharply related to the specific area of skin supplied by the spinal nerve branches; the latter is related to the nerves supplying the midline gut and so such pains are diffusely referred to the midline. They are usually referred to the upper portion of the abdomen for foregut pain, around the umbilicus for midgut pain, and above the pubis for hindgut pain.

4.2.5 Servicing the Gut

The division of the gut tube and peritoneal cavity into their several parts is not merely for descriptive convenience, but fundamental in arrangement. Thus the three central parts of the gut are midline unpaired derivatives of the internal trunk alone. The cranial-most (mouth) and caudal-most (anus) parts are produced in relation to lateral paired components of the external trunk. Further, the suspension of these three gut tube parts into what eventually become the various serous cavities (pleural, cardiac and peritoneal) also has implications more than merely descriptive.

As a result, whereas the arterial, venous, lymphatic and nerve supplies of most other structures of the body involve paired components on each side (e.g. paired thoracic and lumbar arteries, paired azygos veins, paired lymphatic chains, paired spinal nerves), the three parts of the gut tube are supplied by single midline, arteries, veins, lymphatics, and nerve plexuses, in phase with these three sections. And these services are intimately related to the serous membranes in the various body cavities.

There are thus three midline arteries coming from the ventral surface of the main body vessel (the descending aorta). The celiac artery supplies the main parts of the foregut, the superior mesenteric artery the midgut, and the inferior mesenteric artery the hindgut. Associated with these vessels are midline veins and midline lymphatics. The lymphatics pass to the three parts of the gut with the respective branches of the arteries. Likewise, the nerves, autonomics, to the gut tube, are distributed through three midline autonomic plexuses associated with the base of each midline artery. All these service structures pass to the gut from the back of the abdomen and, in those regions where portions of the gut tube are suspended by infoldings of the serous membranes, between the two layers of these infoldings.

In complete contrast to these parts of the gut are the arrangements in the upper-most (mouth and related tubes) and lower most (anus) parts of the alimentary tract. These parts are supplied by paired nerves and vessels from the head (for the oro-facial portion) and from the external trunk (for the perineal portion) and are not involved in folds of serous membranes.

It should be noted that there are other neurovascular elements in the thorax and abdomen that pass to other non-gut structures or derivatives that are paired (e.g. renal system, genital system – see below). These are, are of course, arranged as paired segmental nerves, arteries, veins, lymphatics, and so on. These paired service elements are associated with the paired (bilateral) developmental origins of these particular structures. Because they lie on the dorsal wall of the abdomen, these services are not related to serous membranes (except being covered by them on their ventral surface).

For further information and a more complete description of the serous membranes see later this chapter.

4.2.6 The Respiratory System

A development of the gut tube that starts in the head, the eventual respiratory system, is also important. Very early in the embryo the foregut produces a ventral evagination called the respiratory diverticulum or lung bud. Of course, this lies at first in the putative head. It rapidly lengthens, however, and bifurcates into two channels, the primary lung buds. These give rise to the primordial lungs by a series of further bifurcations (producing the branched tree of what will be the air-conducting passages in the adult).

Near birth the terminal air passages produce the first air sacs (primitive alveoli), and these become the gas exchange portion of the lungs. This lengthening system thus descends from the head past the neck to end up in the thorax, invaginating itself into that part of the serous cavity that, thus, becomes the two pleural cavities. As this structure passes the caudal part of the head it picks up its nerve supply from branches of the vagus nerve, (which therefore, though a cranial nerve, has these long branches that pass down into the thorax in relation to the lungs). Similar branches of the vagus nerve also pass to the developing heart in the thorax, and indeed, into the abdomen serving the foregut and midgut tubes. There are also equivalent long branches from the cervical portion of the developing sympathetic system to the thoracic viscera.

4.2.7 The Serous Cavity

The developing serous cavity surrounds the greater part of these gut and gut-related structures on their ventral and lateral aspects. The vessels and nerves enter the structures between the parietal and visceral layers on their dorsal aspect. Some of the developing structures, for example, the rectum, actually lie dorsally in the abdomen, being covered by the serous membrane only on their ventral surface. Most of the developing structures come to be invaginated into the serous cavity. As a result, they are almost completely covered by serous membrane, save for a dorsal strip where that membrane flows back onto the body wall to becomes the parietal layer. This means that large parts of the gut tube are apparently suspended in the abdominal cavity by a (usually) dorsally located double serous membrane called a mesentery.

It is between the layers of this dorsal mesentery that the various vessels and nerves pass from their main midline dorsal origins to enter the gut in the dorsal midline. These mesenteric structures are quite complex in some regions forming folds and pockets. A few structures (e.g. the ascending and descending colons), originate in this way, being originally mostly covered by the serous membrane, but subsequently come to lie against the posterior wall of the body with serous membrane only on their ventral surfaces.

Finally, what commenced as a single serous cavity gradually becomes more complex. This complexity is associated with the developmental origins and migrations of the precursors of the diaphragm and heart. As described previously, at a very early stage when the embryo is still a flat plate, the most cranial part, far above the origins of the brain, there develops a series of small blood vessels. Even more cranially to these blood vessels arises a mesenchymal mass called the septum transversum. As the head folds ventrally, the septum transversum and the small cluster of blood vessels also fold ventrally and move caudally. As a result, their relative positions gradually become reversed with the septum transversum ending up caudal to the blood vessel clusters. These cellular and vascular masses continue to descend on the ventral aspect of the embryo. As they pass the cervical region they pick up their nerve supply. This accounts for the diaphragm (that develops in part from the septum transversus) being supplied in part by a nerve (the phrenic nerve) that has its segmental roots from lower cervical levels. It also accounts for the heart (the derivative of the cluster of small blood vessels) having its nerve supply from cranial and upper thoracic spinal levels. These structures eventually come to rest about half way down the developing body cavity with the developing heart lying on the developing diaphragm.

The septum transversum thus comes to partition the serous cavity of the trunk into a caudal portion, the abdominal cavity, and a cranial portion, the thoracic cavity. The diaphragm thus separates the serous cavity into thoracic and abdominal parts.

The thoracic portion of the serous cavity is rendered further complex by the formation and descent of the respiratory tubes described above. These descend into the thorax where the two tubes (to become the bronchi) subdivide a number of times, eventually giving rise to progenitors of a pair of lungs. As a result of this, and the coalescence of the many small blood vessels into two blood tubes that eventually become the heart, the thoracic serous cavity becomes tripartite. There are two lateral pleural cavities within which are invaginated separately each developing lung, and one median pericardial cavity into which is invaginated the blood tubes that eventually become the heart.

For a period narrow canals remain joining all three cavities. Later these close completely, thus forming the three totally separate serous sub-cavities in the thorax, two pleural and one pericardial.

Likewise, as the diaphragm increases in size the communications of the developing pleural cavities with the peritoneal cavity are reduced to two small pleuro-peritoneal canals. These, too, gradually pinch off, thus rendering the single peritoneal cavity separate from the two pleural thoracic cavities.

Occasionally one or more of these canals persist into childhood or even adult life. The existence of such variations not only provides adult evidence of the developmental change, but also sets up the possibility of clinical problems (such as an organ herniating into the wrong cavity, fluids or infections moving easily from one cavity to another, and so on).

Again, these cavities do not normally contain anything except a small amount of serous fluid emanating from the lining membranes. The organs that one might think were inside these cavities are, of course, not inside, they have been invaginated into the cavities. As a result, the invaginated organs are covered by the invaginated part of the serous membrane. It is in this way that the serous membranes come to be described as having a parietal layer (lining the body wall), a serous layer (overlying the invaginated organ) and a double 'meso-layer' where the two come together, in the root of the organ.

It is, of course, within the two layers of the 'meso-layer' that blood vessels, nerves, and so on, traverse from the dorsal aspect of the abdomen into the more ventrally located organs. They do not cross the serous cavity. The term 'meso-layer' is used here to describe the general picture; the terms actually used, such as meso-gastrium, meso-colon and so on, are named in relation to the appropriate specific named organ to which the layer is related.

Thus it is useful to realise that the trunk (abdomen or thorax) holds all the structures we have been describing, including the serous cavities. There are no such cavities in the neck. The serous cavities themselves contain nothing save small amount of serous fluids. Thus these cavities and their membranes, lubricated by the serous fluid, form systems that allow the various organs (e.g. lungs, heart, and gut) to move around as they function in the living individual. This arrangement also permits substances that should not be in this space (e.g. blood, pus, foreign objects and so on) to cause medical problems, or to allow organs to slide into the wrong places leading to internal herniae of one kind or another.

4.2.8 Another Portion of the Mesoderm: The Intermediate Mesoderm

Further structures in the interior of the trunk are produced by development of the intermediate mesoderm. This is a paired mesodermal cell aggregation that lies at the dorsal junction of the paraxial and lateral plate mesoderm on each side (see Fig. 4.2 (b)). Its developmental story is complex and relates to the formation of both the urinary and genital systems.

A First Intermediate Mesodermal Structure, Urinary (Renal) System

Some early very cranial components of the urinary system are segmentally arranged, probably under influence from the adjacent head and neck somites. Later parts are non-segmented structures, built in relation to the adjacent, unsegmented lateral plate mesoderm of the thorax and abdomen. In the adult the eventual results are the renal and gonadal systems, and their various ducts and other adnexae.

The beginnings of the renal system are in a pair of nephric ridges in the cranial-most intermediate mesoderm on the dorsal wall of the embryo. It is first evident in the cranial end of the nephric ridge as a series of transient, non-functional, clearly segmental bilateral nephrons (the individual elements of a nephros or 'kidney') hence a 'pronephros'. These are actually located in the neck. They may represent a ghostly vestige of the primitive kidneys (pronephros) that are the primary excretory organs in other very distantly related vertebrates.

The pronephros then regresses and is replaced by a pair of more caudal elongated segmented columnar structures, also differentiated in the nephric ridge: the mesonephros. These lie on each side of the dorsal wall in the thoracic and abdominal regions. They are functional for a period but contain only simple nephrons, the microscopic working parts of a kidney. They, too, represent structures that exist in the adults of other distantly related species.

The mesonephros in turn regresses and is replaced at its caudal-most end (i.e. deep inside the pelvis) by a metanephros which becomes the definitive kidney in humans and other mammals. The metanephros is not segmented and grows in the reverse direction, that is, cranially, cranially from the pelvis into the abdomen. (This is the reverse direction of the growth of almost everything else.) It originates from a bud on the duct (the mesonephric duct, see below) that previously served the mesonephros. All these structures lie very dorsally; the eventual outcome, the metanephros (the kidney proper), is thus on the dorsal wall of the abdomen.

Being associated with excretion, these structures all have ducts. The pronephros displays only very small duct-like structures from each individual nephron (also called a nephrotome). The mesonephros on each side develops a major duct (mesonephric duct, or Wolffian duct) leading from the collected nephrons down to structures in the developing cloacal region. The metanephros, being a bud from the lower end of the mesonephric duct, takes over the distal portion of that duct as its own, eventually the definitive ureter. In the adult urinary system, in the normal situation, the metanephros (kidney) and its part of the mesonephric duct (ureter) are all that persist.

The most distal portions of this system, where the excretory tubes open up on the body surface are produced by interactions in the aforementioned cloacal membrane. This includes the origin of a midline urine-storing structure (the eventual bladder) and, draining the bladder to the exterior, a mid-line duct (the eventual urethra). These structures are formed by a ventral split from the lower portion of the hindgut, hence their midline origin and position. However, this lowest portion, though partially originally hindgut in origin (endoderm) is also differentiated through endodermal/ectodermal contacts in the cloacal plate. These components are also developed in relation to the developing genital system. Accordingly, they are described at the end of the next section.

The nerves and vessels of this entire system are paired bilateral components rather than single midline components of the gut tube and its derivatives.

A Second Intermediate Mesodermal Derivative: The Genital System

The genital system arises in close conjunction with the renal (urinary) system. The lynch-pin of the genital system is the gonad. The paired gonads, one on each side, arise from primordial germs cells that have, at a very early stage, migrated from extra-embryonic tissues. These cells come to lie medial to the paired renal ridges as a pair of genital ridges. Genital ridge mesenchyme forms aggregates of cells, the primitive sex cords, which surround the primordial germs cells. This complex develops in all normal embryos, but after the sixth week it pursues a different pathway in female as compared with male.

At the same time a new set of ducts form just lateral to the already present mesonephric ducts that are associated at this stage with the aforementioned mesonephros. These new ducts are called paramesonephric ducts (Mullerian ducts) and are also present in both female and male.

Thus, at this stage (end of the sixth week of gestation) an undifferentiated gonadal precursor and two sets of ducts are present in both sexes. Although male and female seem indistinguishable at this time, subtle cellular and molecular differences are already present.

From this point on, in the absence of Y chromosomes, female development ensues; the female pattern is, thus, the default situation. Development of the male system only occurs if instigated by factors encoded on the Y chromosome. However, this is by no means the only determining factor in this complex series of events. Subsequent phases of sexual development are controlled not only by the sex chromosomes genes, but also by autosomes coding for hormones, and by other factors. In other words there is a cascade of events leading to the eventual adult structures. Thus femaleness is the basic developmental pathway in embryos of both chromosomal sexes, and this is the pathway that is followed unless maleness is actively initiated. This process can also go wrong resulting in a series of intermediate or extreme developments.

The Female Sex Organs

In the female the primitive sex cords degenerate and the mesothelium of the genital ridge forms secondary sex cords. The cortical cells of these cords then surround the primordial germ cells thus forming a structure with a follicle surrounding sex cells, in other words, a presumptive ovary. Gamete production is started at this stage but is then arrested, and does not further change until puberty, at which point the individual germ cells, now oocytes, resume gametogenesis in response to each monthly surge of various pubertal hormones.

At this early stage, too, the part of the mesonephric ducts adjacent to the developing gonad, and which require testosterone for their maintenance, degenerate on each side. The paramesonephric ducts give rise to the tubes along which female sex cells will migrate at puberty. These are the Fallopian tubes. Further, the union of the tubes in the midline form the midline uterus and most of the vagina. Some of the most superficial portions of the vagina stem from the cloacal plate (which has become divided into an anal sinus and a urogenital sinus). These developments occur not only in the absence of testosterone but also in the absence of mesonephric duct (Mullerian duct) inhibiting molecule (see later) which is partly why the paramesonephric duct persists. Various vestiges of other structures may remain.

The Male Sex Organs

If the sex factor is present on a Y chromosome then maleness commences. The medullary cells in the sex cords then go on to differentiate into cells that will form the eventual testis together with associated structures that, again, complete their development at puberty. The secondary cortical cells (that became the ovarian structures in the female) disappear (see above).

Likewise, again, a set of ducts develops and a set disappears. The mesonephric duct adjacent to the disappearing mesonephros persists, presumably because it is the mesonephric tubules that are linked with the medullary cords (putative testis structures). These tubes go on to become the tubes associated with the passage (eventually) of male germinal cells. They include the small ducts of the testis, and the main duct from the testis. These ducts come together in another structure, again a single midline structure (vide the female) at the base of the developing bladder called the prostate gland. The eventual pathway to the outside of the body is through the urethra, already developing in relation to the urinary system.

Both Sexes

Finally, in both sexes, the gonads descend from their original position in the upper abdomen to much lower positions. The testes descend into the pelvis, thence through the external body wall via a canal in the inguinal region, ending up in the scrotum. This has been guided by a ligament, the gubernaculum, whose caudal attachment is in the scrotal swellings. A similar descent occurs for the ovary under the guidance of a gubernaculum. But the ovary itself descends only onto the lateral pelvic wall, though the gubernaculum passes, as in the male, all the way caudally, through the inguinal region and into the labia majora (labial swellings equivalent to the scrotal swelling in the male).

External Genitalia

Even the early development of the external genitalia is similar in both males and females. In the region of the cloacal membrane a pair of lateral swellings form called cloacal folds. These meet in the ventral midline to form a genital tubercle. At this point we can define the elements of the adult. The genital tubercle goes on to form the glans and shaft of the penis in the male and the clitoris in the female. The anterior parts of the cloacal folds form that part of the penis surrounding the penile urethra in the male, and the labia minora in the female. The tissues lateral to the cloacal folds (called the labioscrotal folds) become the scrotum in the male and the labia majora in the females. Only one structure, the central groove in the urogenital fold, called the urogenital groove, becomes different in each sex; it forms the penile urethra in the male and the lower part of the vagina (vestibule) in the female. Even this is not as different as it appears; all that has to happen in the male is for the ventral surface of the urethra to fail to fuse, and for the urethra therefore to open in the perineum, rather than at the end of the penis, for the situations to appear almost identical in each sex. All of this accounts for the difficulty of sexing an embryo on morphological grounds.

Summary of Urogenital Anatomy

In both sexes, then, the urinary and genital systems have a remarkable confluence. They start as paired structures in relation to the paired intermediate mesodermal masses. They each utilize paired structures and tubes in separate formations. The lateral parts then come together caudally where they use materials from the caudal-most parts of the midline gut tube. They both produce their separate external genitalia from the ventral part of the cloacal region, the dorsal part of which gives rise to the termination of the gut tube, the anus. A wide variety of structural anomalies can occur when any of a number of parts of this complex process fail or become deviant.

4.2.9 Overall Summary

In this way we now have all the primary elements of the external and internal trunk. The limbs, of course, are special components, not only of the external trunk but of one part of the external trunk, the ventral component. Because of their special complexity we treat them in the next chapter.

4.3 The Trunk: Comparative Plans

To what degree is the trunk framework revealed in development also evident from the comparative view? The comparative view indicates just how much of human structure is common among living species today, and how much has been differently evolved in humans. This view helps explain why a much older generation of biologists believed that 'ontogeny recapitulates phylogeny' (i.e. that the successive developmental stages of an individual mirror what is evident from comparing adult individuals of many different species), and that the animals alive today formed a 'ladder of life' (i.e. that they were sequentially related to one another). Though there was considerable usefulness in these ideas in earlier times, it is now clear that they are wrong. The situation is more complex.

First, the developmental programs themselves evolve, so that, though there are similarities in developmental stages across organisms, these are modulated, sometimes very strongly, by the evolution of the developmental sequences themselves. As a result, the sequences of developmental forms are not representative of the adults of so-called 'more primitive' species.

Second, the various species alive today are, themselves, the products of as complex a series of branching mechanisms over evolutionary time as are humans. Certainly, the various species today are not representative stages of human development as was envisaged from that older view of the 'ladder of life'. Obviously, no form living today can have given rise to any other any other present day form.

Nevertheless, we can compare the various species with one another to define in part the differences that have evolved, some due to functional adaptations, and the similarities, some being due to similar but modified underlying blueprints and processes. This allows us to understand to what degree developmental programs are fundamentally similar and to what extent diversions have occurred over evolutionary time.

4.3.1 Chordates but not Vertebrates

The tunicate (urochordate) plan already described is relatively simple. Of course, there are a number of different tunicate species, but a general description can cover them all.

Some adult tunicates are sessile creatures with, apparently, no external trunk. It is as though the creature were entirely a head with branchial basket (or pharynx) perforated by branchial slits (and of course, this is slightly reminiscent of the pharynx within the human head).

However, the free swimming larval form does have an organization a little like an external trunk. It contains a notochord (a structure providing a degree of rigidity and therefore acting like a skeleton, but, of course, not made of bone), a series of about 50 muscle blocks (from cranial to caudal), and a longitudinal nerve net in the tail. The muscle blocks are quite curiously arranged, the sets on one side being half a segment out of phase with those on the other. On contraction they bend the notochord in a rather complex way so that the animal moves (swims) in a rotatory manner. This is completely different from the human arrangement. Yet, like a human, it does evince an axial structure (the notochord) present at a very early stage in the human. It also has a segmental structure, even if different from those in humans.

There is little equivalent to an internal trunk. In the adult, the main food-intake is the basket-like pharynx though there is a very short stomach and intestine but these are present solely in the 'head'. In some non-feeding larval tunicates the gut is not fully differentiated so that there is no anus, though there is a tiny continuation of the relatively very much larger branchial apparatus. As the motile larval form metamorphoses into the sessile adult, the elements in the external 'trunk', the notochord, muscles and nerve net are largely resorbed. In other words, most of the features that define such a creature as a chordate disappear in the adult. There is no specialized excretory organ (no kidney or equivalent) but there is, of course, a gonad, though this produces both egg and sperm cells.

The lamprey (cephalochordate) plan is only slightly more complex. However, the entire animal is free-swimming throughout life. Thus the external trunk, containing a notochord, a dorsal tubular nerve cord, and a set of many muscular segments along the trunk, does not disappear in the adult. The internal trunk, too, is evident as a rather simple tube with a foregut leading from the many-slitted pharynx through a short midgut, to a short hindgut to the exterior through an anus. There is no kidney per se, though there are nephridial tubules, collecting wastes mainly from the head and passing them to the atrium, the final passage where water from the pharynx leaves via the pharyngeal slits. These tubules bear a degree of resemblance to the functionless and transient pronephros in the cervical region of the early human embryo. There is also a rather more distinct functional pronephros just behind the pharynx with its pronephric duct passing down to a cloaca as the excretory organ. However, as the animal grows, nephric elements are added caudally so that a mesonephros becomes evident and the pronephros gradually disappears. Likewise there is a gonad in the form of ridges medial to the kidney structures.

In the young animal the organ is 'hermaphrodite' containing developing eggs and sperm together. Later in life, however, the sexes become determined producing only eggs or sperm. There is much here that resembles the situation in humans.

4.3.2 Chordates and Vertebrates

It is, however, the vertebrate plan that starts to be considerably more similar to the actual situation in that one vertebrate with which we are concerned here, the human.

The External Trunk in Fishes

A general pattern (but by no means identical in every fish) shows a segmented series of intersegmental bony or cartilaginous elements (ribs and vertebrae) that are moved by a segmentally arranged series of muscles supplied by equivalent segmentally arranged spinal nerves. All of these structures show dorsal and ventral divisions (Fig. 4.9) and this is reminiscent of the dorsoventral division that occurs early in human development, as described previously.

The Internal Trunk in Fishes

The internal trunk in fishes contains a gut that is now somewhat more complex. It shows, caudal to the pharynx, an esophagus leading to a true stomach and a 'small' intestine that is truly quite short (but rather large because of the presence inside it of a spiral valve, presumably associated with enzymic mixture of foods with absorption). This leads in turn to a short rectum ending at an anus. There are a liver and pancreas.

These arrangements differ, however, in many fish, and that described above is of fishes that are mainly carnivores. It is thus easy to find, in a particular fish, a system that looks like that in other forms. To attach any special phylogenetic significance to this for other forms is probably largely spurious; the systems show such variability as a result of divergent, parallel and convergent evolutionary developments that anything that can be has been.

The urinary system in fishes usually commences with a functional pronephros but this usually degenerates and is replaced by a mesonephros. A wide variety of reproductive structures and functions have evolved in fishes that bear little resemblance to what is found in mammals. Indeed, these complexities have, in the past, confused attempts to understand the evolutionary situation – too many unsubstantiated links having been created by different investigators.

The Trunk in Many Amphibians

Amphibians, now at least partially land animals, have a rather more complex external trunk. The skeletal elements are intersegmental vertebrae that have

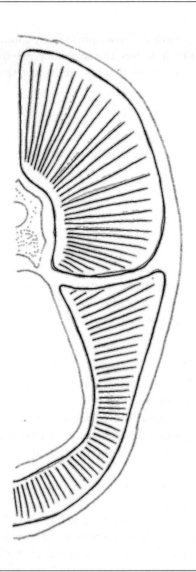

FIGURE 4.9

Cross-section of a fish showing the epaxial and hypaxial components of the segmental muscles.

different detailed architectures related to the different functions in the different spinal regions (neck, trunk and tail). Likewise the segmented muscles are also more complex, being arranged in a series of layers (from dorsal to ventral, and from superficial to deep, varying, of course, in different amphibians, but seen on average in Fig. 4.10). Thus, there is a dorsal longitudinal block of muscles (which therefore extends the spinal column). This seems similar to the dorsal muscle block that develops from the epaxial myotome in the human. There is a much larger group of

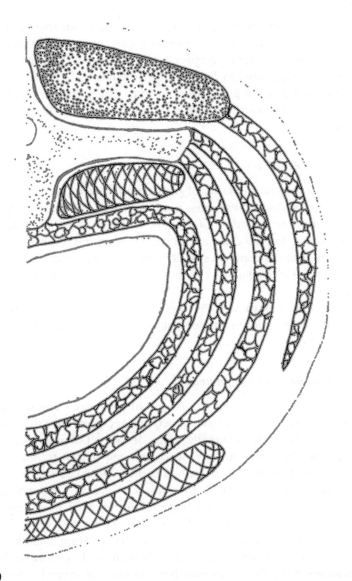

FIGURE 4.10

Cross-section of an amphibian showing (speckled shading) the epaxial muscular block, and various hatched shadings, the complexities of the hypaxial muscles (compare with fish, Fig. 4.9).

ventral muscles like the much larger ventral, hypaxial, myotomal derivatives in the human. These are also organized into a dorsal group of longitudinal muscles on the ventral surface of the spinal column, a similar longitudinal ventral muscle block on the ventral aspect of the trunk, and a series of as many as four or more layers of muscle surrounding the body cavity of the trunk. This is somewhat similar to the developing human – but not the same. These muscles all participate in the complex lateral trunk movements (waves) in swimming movements of the creature when in water, together with equivalent complex lateral waves of the trunk for locomotion when that trunk is supported by limbs while on land.

The internal trunk in amphibia does not evince any particular mechanisms. After all, the vegetative mechanisms vary far more on the basis of diet, whether carnivorous or herbivorous, and so on.

The Trunk in Some Reptiles

This has, again, a general similarity of plan but the details are somewhat different. The vertebrae are clothed by the same fundamental muscle blocks, dorsal and ventral. However, the dorsal muscle block now contains three separate components within itself: medial, intermediate and lateral. The ventral muscle blocks now have their dorsal and ventral components as quite separate, and their divided lateral components may be in as many as four or more layers, with, on occasion, conjoined layers (Fig. 4.11).

The internal trunk, too, though displaying great variations in relation to diet and so on, and a somewhat more complex gut, evinces very little differences from that in the amphibians.

The Trunk in Mammals Including Humans

The picture here is different again, as described above. The dorsal muscles of humans have become elaborated as a superficial set of muscles that are arranged in three groups from medially to laterally, and a deeper set of muscles are also divided into three groups, but these are arranged from superficial to deepest of all tight against the vertebral column, as described earlier (Figs. 4.12 and 4.13).

The increased complexity of these trunk muscles is associated with the very complex sets of movements of the vertebral column allowed in mammals, including humans. However, there are major differences among the various mammals; the bending and twisting movements of the parts of the vertebral columns of kangaroos, cheetahs and whales, are much greater than in humans. In particular the trunk muscles in whales are not only enormously big, but also enormously complex, as befits creatures able to swim, jump and dive in the ways that they can. In contrast, those contortions which the human spine seems able to sustain (e.g. in gymnasts) are actually partly spurious. Though visually large, they are at least partly due to changes in the positions of limbs. Furthermore, though lateral bending (waves) of the vertebral column is still possible in all mammals, including humans, the dorsoventral bending and medio-lateral twisting movements are also extremely important components of trunk movements and of locomotion.

The ventral muscles in mammals are also different in different species (also Figs. 4.12 and 4.13). The longitudinal muscle blocks at the back and front of the trunk are still evident. But they now differ in the different sections of the trunk:

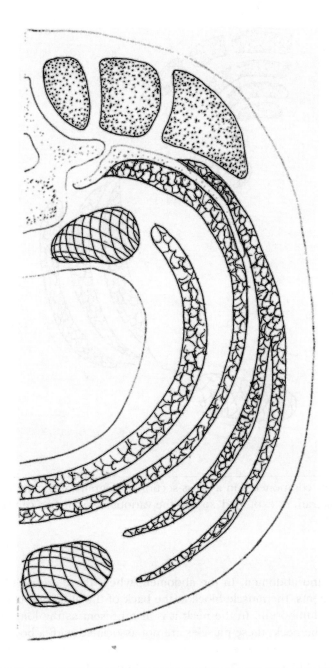

FIGURE 4.11

The equivalent cross-sectional diagram for a reptile showing the various myotomal muscular components. This diagram is 'exploded' so that the various components can be more easily seen.

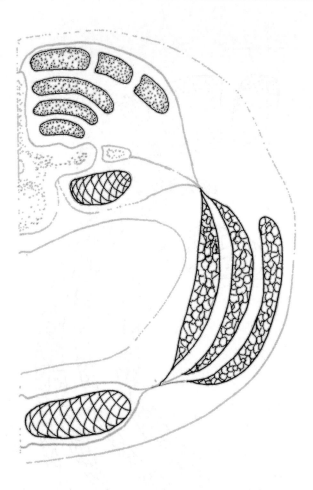

FIGURE 4.12

The equivalent components in a cross-sectional diagram of the abdomen in mammals. This diagram is 'exploded' so that the various components can be more easily seen.

neck, thorax and abdomen. In the abdomen, where there are no ribs free of the vertebral elements, the muscle block at the back of the abdomen is represented as the quadratus lumborum. In the neck it is also present as the long neck muscles. Of course, in the neck, these muscles are not associated with a body wall external to a body cavity. In the thorax with its associated ribs, less movement is possible; and the muscle block at the back is, thus, scarcely evident.

Likewise, the muscular block at the front of the trunk differs between neck, thorax and abdomen. In the abdomen there is a very large and strong rectus abdominis muscle. In the neck this block is also present, but again, not as the exterior wall to a body cavity. In contrast, in the thorax where not only are ribs present, but a rigid

FIGURE 4.13

The equivalent components in a cross-sectional diagram of the thorax in a human; the three layers so evident in the abdomen are still present in the thorax but are divided by each rib; further, each component is very narrow so that it cannot easily be shown in the diagram.

sternum has developed, there is no ventral muscle block (see again Figs. 4.12 and 4.13).

Interestingly, those muscle blocks that are absent in the thorax in relation to the presence of ribs, costal cartilages and sternum, are nevertheless present in a few humans. One of these is a persistent rectus thoracis (see earlier and compare rectus abdominis) muscle. This must be differentiated from a persistent sternal component of the panniculus carnosus muscle (also see earlier).

There are major functional differences in these muscles throughout mammals and these relate to differences in mode of locomotion. In this regard, bipedality in humans is unique, the bipedalities of other creatures, for example, kangaroos, bears, monkeys and apes, being quite different from that in humans.

The evidence that all these variations are due to modulations on a common developmental system is evident when we examine those variations that can be seen in a few bodies. Thus in the back of the human thorax one sometimes finds a few individual small muscular bundles that appear to be elements related to what is left of the muscle block so strongly represented in the abdomen by the quadratus lumborum. Equally, in the thorax, there can be found, in some individuals, muscular remnants (persistent rectus thoracis) on the sternum that seem to represent that part of the ventral-most muscle block generally absent in the thorax. This can be distinguished from another muscle occasionally found on the sternum that is not a derivative of this somatic muscle block, but a variant of a cutaneous muscle layer related to the panniculus carnosus muscle (see above).

Once again, variations among mammals in the internal trunk are far greater than differences between any one mammal and humans. The gut varies enormously in relation to diet. But its main components, foregut, midgut and hindgut, and their subsequent more detailed parts, are generally very similar. The respiratory, urinary and genital systems are also largely as described for humans.

4.3.3 Summary

These comparative evolutionary patterns in the trunk in a wide variety of creatures all strengthen what we have learnt about anatomical patterns in the developmental and functional stories of the earlier sections of this chapter.

CHAPTER FIVE

Building Human Limbs

5.1 The Limbs, from Fetus to Adult

The building of a limb, like the building of a trunk, also depends upon the human 'blueprint' of genes that produce limbs on trunks, and the human 'tool-kit' of molecular factors that determine the discrete positions on the trunk where the human limb fields are generated. The structures of the limbs likewise are related to the diversification of forms living at the same time, and their plans and patterns. They are due to effects of mechanisms during function, and questions of adaptation and optimization. They are also related to the integration between body and brain over control

The Scientific Bases of Human Anatomy, First Edition. Charles Oxnard.
© 2015 John Wiley & Sons, Inc. Published 2015 by John Wiley & Sons, Inc.

time and the importance of controls and communications. The foregoing concepts can then be seen as integrated in evolution over deep time as revealed through the relationships between animal architectures and lifestyles. Together all this can be considered as the science underlying limb structure.

5.1.1 Multi-segmental Origin of Limbs along the Trunk Axis

A first level of organization in the limb relates to the multisegmental origin of each limb, and to its limited disposition along the trunk. We already know that segments are evident from head to tail, but the precise segment numbers, and the precise positions within the segments, at which limbs appear, differ in different animals. The choice is due to the particular homeobox genes and molecular factors involved. Some vertebrates, for example, snakes, do not have any limbs. Even in snakes, however, the developmental blueprints and tool-kits are still present. Appropriate experimental manipulations can induce limbs to form in snakes, and extra pairs of limbs to appear in chicks (see Chapter 2).

5.1.2 Ventro-lateral Origins of Limbs within Trunk Segments

Notwithstanding some variation, limb positions in any given species are fairly constant. The limbs are first evident as small buds on the ventro-lateral aspects of the trunk. Molecules such as retinoic acid (acting through the Hox-gene system) appear to be critical for the initiation of the limb bud. These molecules stem from the more ventrally located notochord and, at later stages, from the ventral components of the peripheral nerves developing at each segment. Thus limb buds are ventral derivatives on the trunk wall; they are not related in any way to the dorsal portion of the trunk (Fig. 5.1(a–c)).

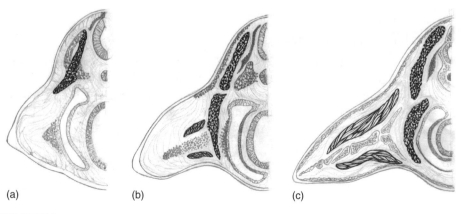

(a) (b) (c)

FIGURE 5.1

(a–c) Components in the early formation of a limb from the ventral part of the trunk.

The materials in the developing limb derive from a series of migrations of materials. These include: migrations of ventral myotomal components of the paraxial mesoderm (as myoblasts, pro-muscle cells) to form limb muscles, of ventral connective tissue components of superficial lateral plate mesoderm to form, eventually, bones, joints, and connective tissues, and growing nerve fibers that, again, following this pattern, stem from only the ventral rami of the spinal nerves.

Whether or not an upper or a lower limb develops from the limb bud appears to be specified by genes that influence molecular factors (such as Tbx4 and Tbx5) at particular segmental levels. For the human upper limb, these levels are the lower cervical and upper thoracic segments, for the human lower limb, the lumbar and sacral segments.

The connective tissue portion of the lateral plate mesoderm requires further description. It has two components. One is called the deep fascia. This surrounds all the other structures of the limb. That is to say, it surrounds the bones and joints, the deep neurovascular components (arteries, veins, lymphatics, and related structures) together with the myotomally derived muscles. Its superficial boundary is called the superficial layer of the deep fascia.

The second part of the lateral plate mesoderm is the superficial fascia. It underlies the epithelium of the skin thus forming the dermis and surrounds the superficial neurovascular elements in the skin of the limb.

This division between superficial fascia and deep fascia is not just some semantic point, but a fundamental architectural arrangement existing not only in the limbs but, as already described, throughout the body.

5.1.3 Multi-segmental Origin 'Hidden' by Compression of Segments

As the limb develops, its initial origination from a number of specific body segments becomes somewhat obscured by compression of segments. Though fins (limb equivalents) in some aquatic vertebrates have a structure that indicates that they are obviously derived from many segments, limbs in tetrapods (except at their distal ends, the hands and feet) look as though they are not segmental structures.

That they are actually derived from several segments is evident not only from study of the various developmental mechanisms and processes, but also by the distributions in the cranio-caudal series, of the segmental dermal, muscular and skeletal (connective tissues and joints as well as bones) components of the limbs. These are not only supplied by the cranio-caudal series of ventral components of the segmental (spinal) nerves to the limb, but are actually influenced by them almost exactly as they would be were they separate segments like the trunk. Thus absence (experimental or pathological) of specific spinal nerves may produce absence of specific muscle, bone and joints.

Though the nerve supply of each limb clearly derives from several specific ventral trunk segments, it becomes compressed into a nerve plexus as it passes through the root of the limb. Nerve plexuses – the brachial plexus in the upper limb, and lumbo-sacral plexus in the lower limb – are structures that are formed by the conjunction of the several individual ventral rami of the relevant spinal nerves. Out of these plexuses come the individual limb nerves passing peripherally into the free limb proper. Notwithstanding the apparent coalescence of segments in the plexus, that they still exist in their segmental arrangement is clear from various pieces of evidence.

Thus, their segmental continuity can be determined anatomically by dissecting individual nerve fibers in a retrograde manner from the peripheral nerve through the complexities of the plexus into specific ventral rami. This can also be detected by inferences from developmental experiments, physiological effects, pathological defects, and studies of genetic syndromes.

5.1.4 Specific Segmental Arrangements

Specific segmental arrangements in limb structure are evident throughout all the limb components.

Most superficially, the dermis of the limb stems from the ventral superficial (or peripheral) mesoderm (like the superficial fascia) of the trunk. This is a layer that is distinct from the deep mesodermal derivatives (muscles, bones, joints all bounded by the deep fascia) of the trunk. Not originally segmented, this mesoderm is influenced by the same developmental factors (sonic hedgehog) and neural influences that influence segmentation in the trunk. As a result, this mesoderm, and the dermis that results from it, comes to have a nerve supply in parallel (see below) with that to the limb myotomes (and the eventual muscles) and the limb lateral plate mesoderm (and the eventual bones, joints and deep connective tissue elements). The nerves supplying the dermis are the cutaneous branches of the specific nerves. For the upper limb this includes nerve fibers from the ventral trunk segments C3 or C4 to T3 or T4, and for the lower limb, ventral trunk segments T12 to S5.

Thus, in the upper limb (Fig. 5.2(a)), the tip of the shoulder (the cranial-most part) is supplied by C3, C4 and C5, and the inner aspect of the arm (most caudal) by T1, T2 and T3. The intermediate portion of the upper limb (arm, forearm and hand) is supplied by the intermediate segments C6, C7, C8 and T1. In many expositions these skin areas are shown with boundaries. Because of the amount of overlap, and especially because of differences among individuals, it is best to show only approximate regions.

The same *caveats* apply to the lower limb (not figured). Thus, in the lower limb, the groin (cranial) is supplied by T12 and L1, the natal cleft and surrounding lower limb skin (caudal) by S3 to S5. The rest of the limb (thigh, leg and foot) is supplied by the intermediate segments L2, L3, L4, S1 and S2.

Information of this type has been used to prepare what can be called 'dermatome maps' (the word dermatome here being used in an anatomico-functional sense, different from the usual embryological meaning). Though these maps really have fuzzy boundaries (because of overlap between segments) they are, nevertheless, useful in providing underlying structure and clinical significance to what might otherwise seem quite confusing patterns (Fig. 5.2(a)).

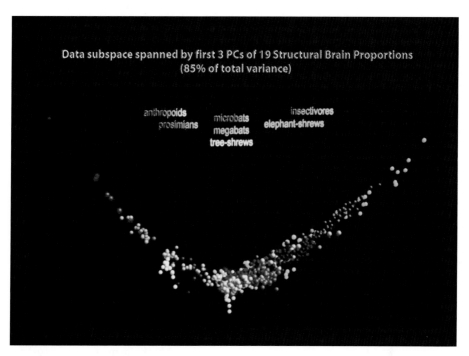

FIGURE 2.5

Analysis of brain sizes separates some different species of mammals. Primates (red and yellow dots diagonally upwards to the left) are organized completely differently from insectivores (various shades of green diagonally upwards to the right) and bats (various shades of blue horizontally organized at the base of the diagram).

The Scientific Bases of Human Anatomy, First Edition. Charles Oxnard.
© 2015 John Wiley & Sons, Inc. Published 2015 by John Wiley & Sons, Inc.

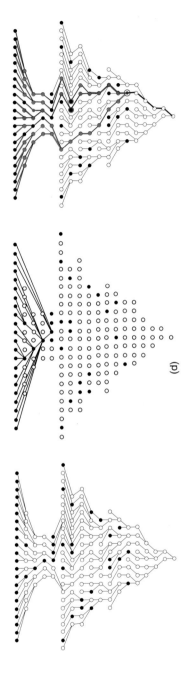

(d)

FIGURE 2.6

A model (d) containing a bottle neck. The common ancestral species as defined by the fossils alone lies only at about the level of the bottle neck. The real common ancestral species is many generation further back and results from a number of different species lineages (as shown by the lines in red, green and orange). In other words, looking at fossil data is rather likely to give incorrect species lineages, and incorrect times by factors of 3 or more.

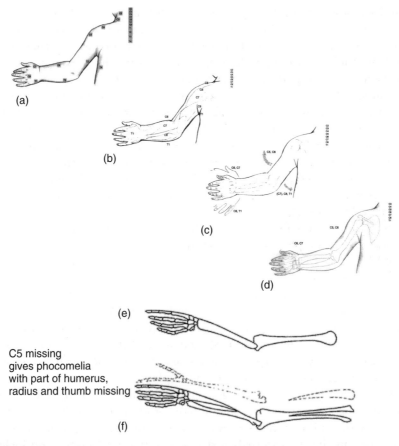

FIGURE 5.2

(a) Approximate upper limb areas (dermatomes) supplied by numbered ventral primary rami of cranial nerves. (b) Approximate upper limb muscle blocks (myotomes) supplied by numbered ventral primary rami of cranial nerves. (c) Approximate upper limb movements ("motor-o-tomes") produced by numbered ventral primary rami of cranial nerves. (d) Approximate upper limb bone components (sclerotomes) controlled by C5 and C6 primary rami of cranial nerves. (e) Bony elements related to C6. (f) Effects of a single sclerotome (C5) deficit in phocomelia.

A similar arrangement, but involving fewer segments, applies to the nerve supplies of the muscles; in this case, however, the nerves are the muscular nerve branches directed to the migrating myotome elements. Thus, in the upper limb, the muscular nerves are limited to segments C5 to T1, in the lower limb to segments L2 to S2. Again, the arrangement relates broadly to the cranio-caudal structure of the limb. For example, in the upper limb the most cranial shoulder muscles (e.g. deltoid) are supplied mainly by C5 and C6. The most caudal shoulder muscles (e.g. latissimus dorsi and the lower portion of pectoralis major) are supplied mainly by C7, C8 and T1. Thus, as with the skin, 'myotome' maps (the word myotome again being used in a sense different from that in development) can be plotted. They, again, are extremely useful in providing underlying structure and clinical significance to what might otherwise seem quite confusing patterns (Fig. 5.2(b)).

Distributions like these mean of course, that the movements which the muscles produce are also channelled by particular segments. Thus, that movement of arm at the shoulder called abduction, which is produced largely by the action of deltoid and related muscles, is controlled by the nerve segments supplying these muscles: that is, segments C5 and C6. That movement of the arm at the shoulder called adduction, which is produced largely by the action of latissimus dorsi and the lower portion of pectoralis major, the main adductors of the shoulder, is likewise controlled by nerve segments, in this case C7, C8 and T1.

Similar concepts apply in the lower limb. One major cranial muscle of the lower limb (iliopsoas) is supplied by the cranial segments L2-L4. Hence hip flexion is controlled by L2-L4. The main caudal muscles (hamstrings) are supplied by the caudal segments S1-S3. Hence hip extension is controlled by segments S1-S3.

This means, yet again, that 'movement' maps can be drawn that help to convert the feat of memory into derivative understanding (Fig. 5.2(c)). In other words, we can think of the muscular branches of spinal nerve segments as not only supplying muscles, but supplying movements. Should we call these maps 'motor-o-tomes'? This parallels the way in which the cutaneous branches of spinal nerve segments not only supply skin areas, but also sensation areas.

Finally, though it is less often recognized, similar concepts also apply to the deep derivatives of the lateral plate mesoderm in the limbs: that is to all the eventual connective tissue (including bone) structures. As explained above, these are bounded, under the dermis, by the external-most layers of the deep fascia, and include the bones and their linking joints, and the complete web of deep fascia that surrounds all deep limb structures. Thus, these connective tissue derivatives (e.g. not only fascial sheets but their contained bones and joints) also relate to the specific spinal nerve segments and their component nerve fibers.

This is fairly obvious in the hands and feet, where the presence of several elements arranged cranio-caudally (the digits) provide clear separations that look like separate segments. But it also exists in the more proximal parts of the limbs (where there seems no evidence of separate segments). For example, in the upper limb, cranial-most segment C5, obviously related to the production of the first one and a half cranial digits (thumb and half of first finger) is also related more proximally to the production of the most cranial portion of the radius, and even more proximally still to the most cranial aspect of the upper humerus and the supraspinous fossa of the scapula (Fig. 5.2(d)). Conversely caudal-most segment C8, clearly related to the production of the caudal-most two digits of the hand (the fourth and fifth) is also related to the production of most of the ulna (caudal in position) and the medial aspect of the lower humerus and possibly part of the clavicle (also both caudal in position).

Approximately similar arrangements are evident on the lower limb where the cranial elements of the limb (big toe obviously, but also some of the tibia and femoral head and neck) are related to L3 and L4, and the caudal-most elements (obviously little toe, but also part of the fibula, femur and even sacrum) are related to L5 to S2.

This means, yet again, that equivalent 'connective tissue' or 'sclerotome' maps can be generated. These have proven most useful not only in helping understanding of normal anatomy (Fig. 5.2(d)), but also in elucidating the skeletal abnormalities that result from developmental problems (such as the thalidomide abnormalities, Fig. 5.2 (e, f)).

In other words, we can understand the segmental arrangements of the ventral branches of spinal nerve segments to limbs as supplying dermis **and** sensation, muscles **and** movements, and even bones **and** mechanics.

5.1.5 Dorso-ventral Patterns within Limbs

While remembering that the limb is a ventral derivative of the trunk, we can also recognize a secondary level of patterning in limbs.

One of these separates the ventral limb into its own internal dorsal and ventral components. This is very clearly obvious in the limb nerve supply. Thus the ventral primary rami (the ventral components of each trunk spinal nerve) are, in the limb, secondarily partitioned into limb dorsal and ventral divisions. It is likely that this is the result of further developmental factors (the Wnt family of proteins) also as modified by Homeobox gene actions, though, as yet, these are less well defined.

Therefore the cutaneous nerve supply of the limb is equivalently organized (of course with considerable overlap of segments, as in the trunk previously discussed). Thus nerve branches from the dorsal divisions of the ventral primary rami from all segments, C3-T3, and T12-S5, supply the dorsal aspect of the respective limbs. Likewise the ventral divisions of the same segments supply the ventral aspect of the limbs. These secondary dorsal and ventral divisions derive mainly from ventral trunk nerves that have already passed through the segmentally compressed structures, the limb plexuses.

This also results in the muscular nerve branches (and hence, also, the muscles that they supply) being likewise arranged in a secondary dorsal and ventral manner. The dorsal muscles are supplied by dorsal divisions of all the ventral rami (but a more restricted number of segments than the skin) C5-T1, and L2-S3, and the ventral muscles by the same restricted numbered ventral divisions.

Thus, though the limb is a purely ventral structure with reference to its derivation as an outgrowth of the trunk, it has its own internal dorsal and ventral organization of nerves, in relation to skin and sensation, muscles and movements, and connective tissue components. All of this, further, makes understanding the anatomy of the limb explicit; it removes much of the element of the rote learning of limb structure.

5.1.6 Axial Patterns within Limbs

A further arrangement: from proximal to distal, is also evident for the limb skeleton. The portion of the limb bones, joints and connective tissues that remain within the trunk are differently generated than in the free limb. Thus, the ventral sclerotome, which, in a generalized trunk region, gives rise to the ventral portion of the vertebrae and ribs, gives rise at trunk levels in the limbs to the limb girdles alone. The dorsal trunk sclerotome, of course, does not contribute to the girdles even though the parts of the girdles (e.g. scapula and ilium respectively) are very dorsally placed in the body.

The Limb Girdles

The limb girdles, like the free limb components are also organized into their own internal dorsal and ventral components.

Thus, at the shoulder, most of the scapula (the scapular blade and part of the glenoid cavity) is a dorsal skeletal component of the limb. From it many of the dorsal limb muscles take origin (except for those dorsal limb muscles that have migrated proximally so that they actually arise from the dorsal aspect of the trunk skeleton, e.g. latissimus dorsi).

Likewise at the shoulder, part of the clavicle, and also part of the glenoid and the coracoid portion of the scapula are, in contrast, ventral limb derivatives. From these skeletal components many of the ventral limb muscles take origin (e.g. coracobrachialis muscle), again except for those ventral limb muscles that have migrated proximally onto the ventral trunk skeleton, (e.g. the various pectoral muscles).

Likewise in the lower limb girdle, the hip, lying wholly within the ventral compartment of the trunk, has an iliac part that is embryologically dorsal in the limb. From two surfaces of this originate the dorsal limb muscles (e.g. iliacus and the gluteal muscles). The pubic and ischial parts of the hip bone are embryologically ventral and from these parts come the ventral muscles (e.g. adductors and hamstrings). (The ischial part looks as though it is dorsal in position, and it is certainly located as 'posterior' in medical anatomy, but it is truly a ventral portion of the girdle. It merely appears to be dorsal in the adult because of rotations of the limb that occur during development, see later.)

The only dorsal muscle of the lower limb that has its proximal attachment on the trunk is the psoas muscle group. But, of course, this group has no effect on the trunk/pelvis junction because, unlike the situation in the upper limb, that junction is a fibrous joint permitting very little movement. The extra proximal attachment of this muscle block on the trunk seems to provide merely additional origin for the function of this muscle at the hip (though it may have minor effects on the lumbar vertebral column). Likewise, though one ventral derivative has its proximal attachment on the trunk (piriformis from the sacrum) it has no effect on the trunk pelvis junction. The other ventral component of the lower limb that arises from the trunk is that associated with the distal tail, to the coccyx (levator ani) and to a movable tail (caudofemoralis). But humans which lack an external tail, and which therefore do not have a caudofemoralis as named, do have the component that attaches to the coccyx: the levator ani muscle. Its function, as its name implies, has come to have nothing to do with the lower limb. Thus even when a movable joint between trunk and hip is absent, as in humans, muscles that are present in other forms are still represented in humans but with new functions.

The Free Limbs

In the free limb the skeletal elements are derived from non-segmented portions of the lateral plate mesoderm. As a result they do not display an obvious dorsal and ventral patterning, but are generally aligned as components along the central axis (axial) of the limb. This difference, between muscular and skeletal components, and between girdle and limb bones, is related to the differences in embryological origination.

Thus limb girdles arise as a result of influence by segmented somitic material (influenced by molecular factors relating to segmentation and dorso-ventral subdivision of muscles) and are, therefore, similarly organized as the muscles.

The free limb bones arise from lateral plate mesoderm lying in the limbs (influenced by molecular factors, possibly fibroblast growth factors, involved in producing connective tissues). As discussed in the previous section, they have elements that can be identified non-anatomically as stemming from segmental influences relating to muscles and nerves. But they remain axial in position and are not organized dorsally and ventrally like the muscles.

In the cheiridia (hands and feet) there is a further distinction in that, though the bones are axial in position, they are axial in relation to a return to a cranio-caudal segmentation, the thumb and big toe being cranial, the little finger and little toe being caudal (see above). This applies not only to bones which are easier to describe, but also to all the connective tissues that enclose bones (periosteal membranes), surround muscles (fascial sheets) and form ligaments, tendons, and joint capsules (all connective tissue structures).

The Deep Service Structures in Limbs: Arteries, Veins and Lymphatics

These concepts also apply to all those other structures that develop within the deep connective tissue of the lateral plate mesoderm of the limb. Thus the main arteries in the deep compartments of the limb (i.e. deep to the outer investing deep fascia) are, like the bones and joints, initially organized axially. Thus the first artery of each limb is the central artery. As development proceeds, most of the central artery becomes very small or even disappears.

Thus, in the upper limb only a part of what seems to be the main artery to the arm, a portion of the axillary artery and a portion of its continuation as the brachial artery represent the more proximal remnant of the original central artery. Likewise

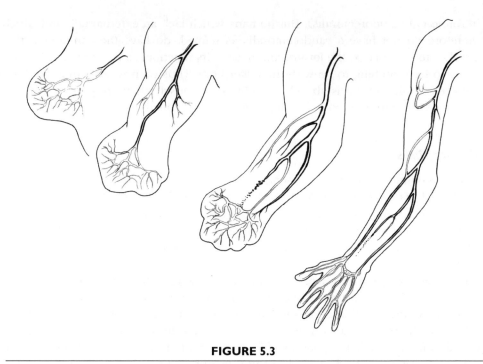

FIGURE 5.3

Arteries of the arm.

in the forearm, only a relatively small artery, the anterior interosseous artery represents the remnant of the central artery more distally. The main vessels of the adult arm and forearm, most of the brachial, radial and ulnar arteries, though much the larger in the adult, are actually secondary branches of that original axial artery (Fig. 5.3).

Likewise in the lower limb, the original central artery is only represented by small portions of the gluteal artery at the hip, the popliteal artery at the knee and the peroneal artery in the leg. Most of the main vessels of the adult (the femoral artery in the thigh, and the anterior and posterior tibial arteries in the leg) are secondary branches of that original central artery (Fig. 5.4).

Similar patterns apply to the veins and lymphatics that accompany these deep arteries (Fig. 5.5). Indeed the veins, usually more than a single vessel, that pass with the main deep arteries are called 'venae comitantes' (which means accompanying veins).

The Superficial Service Structures of Limbs

As in the trunk, those structures that lie superficially, that is, that develop within the superficial fascial layer supporting the dermis, are not organized axially, but cranially and caudally. They develop from the beginning as cranial and caudal

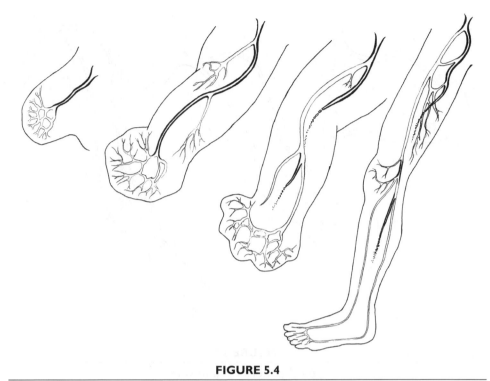

FIGURE 5.4

Arteries of the leg.

components lying on the cranial and caudal borders of the limb, respectively. This also means that they run along the boundaries between the dorsal and ventral components.

In the upper limb the veins are the cephalic (i.e. cranial – the name is descriptive of the concept) and the basilic (i.e. caudal); in the lower limb the veins are the long saphenous (cranial) and short saphenous (caudal). The positions of these veins indicate the positions of these two embryological borders of the limbs. The positions in the adult reflect the twist that occurs in the limbs during development (see elsewhere).

The equivalent superficial lymphatics also lie on the cranial and caudal limb margins. They drain first into superficial lymph nodes. These are located, in the upper limb, at the elbow (cubital or supratrochlear nodes) and at the shoulder (deltopectoral nodes). In the lower limb they lie at the back of the knee (popliteal fossa: popliteal nodes) and in the groin (inguinal region: superficial inguinal nodes).

An important element of these superficial veins and lymphatics relates to the anastomotic connections that they make with their deep equivalents. There are, thus, only relatively few connections between the superficial and deep venous and lymphatic systems along the length of the limbs. Particularly important are the regions, cubital fossa and axilla in the upper limb, popliteal fossa and inguinal region in the

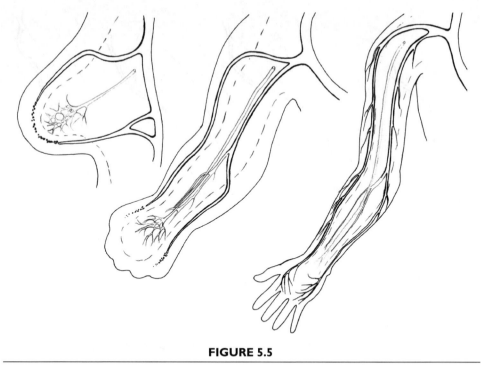

FIGURE 5.5

Typical pattern of limb veins.

lower limb, all where the limb flexures occur. It is in these flexures that the most important links between superficial and deep systems exist, and also where many of the servicing elements (the various veins and lymphatics) lie close together.

The main importance of these relationships is clinical, in relation to clinical examinations, traumas, surgeries and pathologies of veins (e.g. venous thromboses and venous varicosities) and lymphatics (e.g. lymphangitides, lymphedema, lymph node swellings, and so on).

Dorsal and Ventral Terminology, a Caveat

The terms 'dorsal' and 'ventral' occur everywhere in the body and were more fully described in Chapter 3. However, in the limbs, there are very special relationships in these terms that stem from developmental changes, 'twists', that are peculiar to the limbs.

Thus, the skin areas and muscle blocks though properly identified as dorsal and ventral, are also often called 'extensor' and 'flexor' surface areas and muscle blocks, respectively. This is because, in the early embryo, those are the positions (skin) and movements (muscles) relating to the parts.

Later in the fetus, and particularly postnatally, some of the actual positions of the skin areas and muscle functions become changed because of rotations that occur.

These rotations are complex; they reflect several different sets of changes, some developmental, some functional and some evolutionary. They can, moreover, be confused. They produce differences between the embryonic positions and embryonic functional blocks and the actual adult positions and adult functions. This applies especially to skin and sensation, to muscles and movements, and to nerves and reflexes.

The Upper Limb

In the upper limb in the early fetus, the progenitors of certain ventral muscles crossing the shoulder and the wrist are clearly evident in a limb in its very early position. They look out of the page (ventral view) with the rudimentary thumb pointing cranially (it is on the cranial border of the limb (Fig. 5.6).

As the fetus develops the proximal segment comes to show a medial rotation. The single ventral muscle block becomes evident as several pectoral muscles (though only pectoralis major is figured in the diagram). However, a dorsal muscle, a part of deltoid but previously hidden because it was on the dorsal side, comes into view (Fig. 5.7(a)).

As the fetus reaches towards neonatal dimensions with further increase in size and complexity, a similar twist of the distal limb segment starts to be more clearly evident. As a result, the distal ventral muscle component (for example, progenitors of finger flexors, is rotated even further medically and can only be partly seen in this view. The dorsal muscles such as the finger flexors can now be clearly seen as it is the dorsal aspect of the hand that is now visible (Fig. 5.7(b)).

Finally, when we then examine what happens after birth, especially with the medical definition of the anatomical position, the forearm is now rotated laterally so that the thumb lies on the lateral aspects of the limb, the flexors of the fingers are now visible and the extensors are now only partially evident, peeping around the external border of the limb. Of course this is not a developmental twist, but is permitted because the distal segment, with its two bones, radius and ulna, are capable of supination. In the anatomical position the limb is described in this supine position (Fig. 5.7(c)).

The Lower Limb

A somewhat similar series of changes occur in the lower limb (Fig. 5.8(a–d)). The muscle components shown here in Fig. 5.8(a) are the ventral adductors and hamstrings at the hip and the dorsal gastrocnemius elements at the ankle, both evident on the front of the limb.

As the twist occurs, the adductors and hamstrings start to move towards the internal surface and the quadriceps femoris starts to appear on the lateral aspect from its original position on the dorsal aspect of the limb (Fig. 5.8(b)).

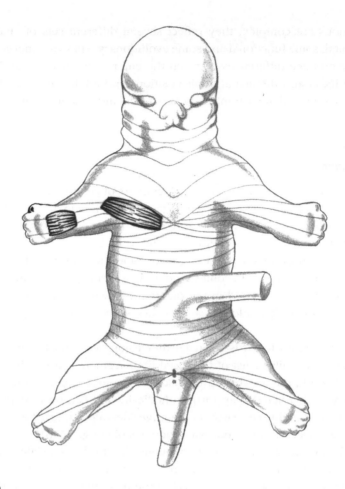

FIGURE 5.6

En face view of ventral aspect of the early fetus showing initial position of upper limb bud and two ventral muscles blocks in the arm (pectorals) and forearm (wrist and finger flexors) respectively; see text.

At the next stage (Fig. 5.8(c)) the dorsal muscle in the leg (a dorsiflexor muscle of the ankle) starts to appear from around the lateral side. Both segments of the lower limb have carried out the same internal rotation as we showed in the upper limb.

Finally, however, we see the situation in the adult (Fig. 5.8(d)). Here, in contrast to the situation in the upper limb, a reverse, lateral, rotation cannot occur. The tibia and fibula do not allow pronation and supination. Accordingly, the adult lower limb remains in this medially rotated position with the muscles on the front of the thigh being the developmentally dorsal muscles, the quadriceps femoris, and the ventral adductors muscles being restricted to the medial aspect of the thigh and the other ventral muscles (hamstrings) being located totally on the back of the thigh. In the

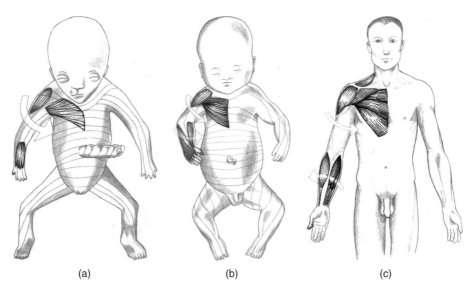

(a)   (b)   (c)

FIGURE 5.7

(a–c) En face view of upper limb bud and positions of same ventral muscle blocks as in Fig. 5.6 at subsequently increasing ages; see text.

same manner, in the leg, the various dorsal (dorsi-flexor muscles of the foot) are found on the front of the leg, the ventral (plantar flexor muscles, gastrocnemius and others are found on the back of the leg.

5.1.7 Superficial/Deep Divisions and Proximo/Distal Migrations

There are yet other levels of organization. They affect primarily the muscles and movements and are related to precise functional factors in the free-living creature (the baby, child, adolescent and adult). They involve superficial/deep arrangements, and proximo/distal migrations. They are nevertheless evident fairly early in the fetus.

Superficial versus Deep Divisions

Thus, at each limb joint, the initial main muscle block, whether dorsal or ventral, becomes divided into superficial and deep components. This is so fundamental an arrangement of animal limbs that it is likely that it is handled by yet other factors in the cascade of processes started by the Hox genes as they produce a limb. However, these mechanisms, if they exist, have not yet been identified. The evidence from anatomy predicts that they are present, and provides hypotheses that may eventually be tested by developmental biologists.

Thus, at the shoulder the set of superficial muscles that develop (and which are attached further from joints) tend to be the ones that produce large movements of

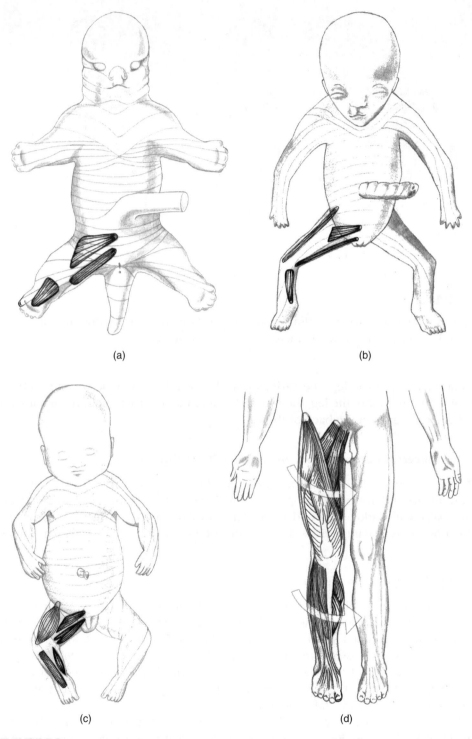

(a)

(b)

(c)

(d)

FIGURE 5.8

(a–d) En face views of lower limb at increasing ages showing effects of limb rotation upon ventral and dorsal muscle blocks respectively; see text.

limbs. These include, for example, latissimus dorsi (dorsal) and pectoralis major (ventral) both being superficial. The set of deep muscles (attached closer to joints) tend to be more related to maintaining joint stability; examples are subscapularis (dorsal) and pectoralis minor and subclavius (ventral), both being deep.

Hip equivalents include the more superficial gluteus maximus (dorsal) and the longer adductors (ventral), and the deeper lesser gluteal muscles (dorsal) and the obturator muscles (ventral).

Similar superficial/deep relationships are evident at all the other main joints in the limbs.

Proximal versus Distal Migrations

Likewise, at each joint, there are a series of proximal and distal migrations of muscles during development, particularly of the superficial muscles.

Thus the proximal ends of the superficial muscles tend to migrate yet further proximally. As a result, the superficial muscles sometimes come to cross the joint more proximal than the one at which they first developed.

For example, at the shoulder the latissimus dorsi (limb: dorsal, superficial) not only crosses the shoulder joint but has migrated to the trunk midline dorsally, therefore acting on the scapula/trunk 'joint'. This is not, of course, a true synovial joint but nevertheless an anatomical arrangement permitting movement of the scapula upon the trunk. It thus comes to be even more dorsal in position than the true dorsal trunk muscles (the spinal extensors).

For example, again, the pectoralis major (limb: ventral, superficial) crosses not only the shoulder joint, but also has migrated to the midline of the trunk ventrally. It therefore also affects movement of the girdle upon the trunk, as well as the movements of the shoulder joint proper. It likewise comes to be even more ventral than the true ventral trunk muscles (e.g. the intercostals).

Similar arrangements occur in the lower limb, for example, gluteus maximus, dorsal, and psoas major, ventral cross not only the hip joint proper but also the joints between the hip and the vertebral column. In the lower limb, however, because the pelvis is fixed to the vertebral column in humans, there is no vertebral-pelvic movement to be considered. What is achieved by the central migration is a larger surface of origin of these large muscles.

A second effect is a migration mechanism that places muscles more distally into the limb. Thus, the distal ends of many superficial muscles may be drawn further distally into the limb, thus crossing a joint below that at which they first develop. This particularly occurs in many of those muscles that are in the standard description of the anatomy of muscles in humans. For example, the longest of the adductors of the hip, adductor gracilis (ventral), descends below the knee and thus acts on both hip and knee. A component of the gluteal muscular mass, the tensor fascia lata is attached, through the fascia lata into the leg, and so not only helps move the hip, but acts on the knee.

Variations

The proximal and distal muscular migrations just enumerated relate to the many variations that are found in some humans. At the shoulder, for example such variants include the chondroepitrochlearis muscle, the various 'extra' heads of the coracobrachialis muscle, the dorsoepitrochlearis muscle, and some of the additional 'heads' of the 'triceps' muscle. Though these are all derivatives of 'shoulder to arm' muscles, many of their fibers pass below the elbow joint, and thus into the forearm.

Similar examples are present in the lower limb; for example, the extension distally of the hamstring muscles so that they also influence the knee as well as the hip. However (as in the upper limb) some components of these muscles are only present occasionally, for example, an extension of the hamstring muscles which, though thigh muscles, not only cross the knee joint, but may have occasional extensions well down the leg, even into the Achilles tendon crossing the ankle joint.

Many of these variations are found in only some individual humans, sometimes in very few individuals indeed, but many reflect the standard arrangement in non-human primates. Their occasional existence in humans is, therefore, evidence of pattern in humans that has undergone modification from the time of the evolutionary separation of humans from some pre-human/pre-ape stock.

Such variations need no longer be thought of as 'anatomical stamp collecting' which is the way they tended to be treated in some of the literature of the last century. Interestingly enough, however, in the century before that, many of the even older comparative anatomists knew very well that they were looking at pointers to embryological and evolutionary phenomena, even though they had almost no detailed knowledge of the impact of development upon evolution. This is a matter that we will go into in more detail later.

The overall functional result of these proximal and distal extensions of superficial muscles is that many of them cross two or more joints rather than one. They are, therefore, bi-articular (or multi-articular) muscles. They thus function in many different movements. They can also produce movements through their own elasticity, especially the elasticity of their usually long tendons, even when not actively contracting. This is not generally the case for deep muscles usually crossing only one joint, that is, uni-articular muscles. They usually do not have long tendons, but mainly fleshy attachments. Their functions mostly relate to single joint movements, especially maintenance of joint stability.

These two processes: proximo-distal arrangement and dorso-ventral patterning are very specifically produced by molecular mechanisms that have been long known in development.

Distal Limb Elements, Hands and Feet: Re-emergence of Multi-segmental Patterns

As we have seen, the entire limb has a multisegmental origin. Though this is quite evident in the details of skin, muscles and nerves, it is largely hidden in the bones and joints of the main portions of limbs.

The girdles (shoulder and pelvis, respectively), the first free segments (humerus – arm bone, and femur – thigh bone), and the second free segments (radius and ulna – forearm bones, and tibia and fibula – leg bones) seem not to show cranio-caudal segmentation. Only the girdle shows dorso-ventral patterning; the free-limb bones are axially located. As a result, the segmented and dorsoventrally patterned muscles, skin areas and nerves are arranged around the apparently non-segmented central (axial) skeletal elements. The study of the developmental segmental components of these parts has shown us that segmental components are actually present not, however, as discrete anatomical components but as molecular determinants. In contrast, when we come to the most distal elements of the limbs, overt segmentation re-emerges.

Thus in the limb periphery (in the upper limb the hand, and in the lower limb, the foot), the digits develop as a cranio-caudal array. The thumb and big toe are cranial-most elements, the little finger and little toe, are caudal-most, and the other digits are intermediate. Thus the pollex (thumb) together with part of the radius and the radial part of the humerus relate to the 5th cervical segment, the little finger, together with part of the ulna and the caudal part of the humerus, to the 8th cervical segment. Likewise the big toe hallux (big toe) together with parts of the tibia and femur relate to the 5th lumbar segment, the little toe, together with parts of the fibula and femur relate to the 2nd sacral segment. Segmental arrangements seem to have re-emerged.

5.2 Limbs across the Vertebrates

5.2.1 'Body-Plans' and 'Diversification'

As in other regions of the body, there are, alongside the developmental bases of limb structure, understandings that stem from comparisons of animals with one another. Many principals of limb organization can be obtained through comparative anatomy. As in other anatomical regions, a vertebrate 'body-plan' can be identified by examining 'patterns of diversification' within vertebrates. This provides those fundamental elements of anatomy that are common to most vertebrates but whose variations also separate many vertebrates.

Likewise, more detailed 'body-plans' can be worked out for mammals as a class within vertebrates. In turn, an even more detailed 'body-plan or pattern' can be seen for the order primates as an order within mammals. The anatomy that this last set of comparisons presents is even more descriptive of humans. Differences of humans, as an individual species, from the other primate species are really only quite small. Finally, though in general terms humans are closely similar to other primates, especially to the African great apes, there are certain specific differences in certain anatomical regions that are descriptive only of humans. Many of these relate, of course, to some of the special adaptations of humans in which they differ from other primates. We will pay attention to these features later.

The existence of the concept of the 'body-plan' stems from the hereditary information carried forward yet also changing over the generations during the processes of evolution. As a result, adapting already existing body-plans to new environmental challenges produces new 'patterns of diversification'. This is an easier developmental task than evolving a totally novel structure (something not existing in the plan). For example, vertebrate pectoral fins (e.g. fishes), wings (e.g. birds) and arms (e.g. humans) are not totally new structures in their respective forms. They are modifications of the components of initial structures developing from ventral components of the cranial end of the vertebrate trunk. In the same way, within limbs, shoulders, arms, forearms and hands, are all structures that develop from the upper limb components of a terrestrial vertebrate 'limb-plan'. The genes and molecules responsible for all of these structures (mentioned as examples) exist even in animals that do not have limbs. As a result, experimental manipulations can induce limbs in species (snakes) normally lacking them, or produce limbs at regions of the body (e.g. mid-trunk) where they do not normally exist.

As an introduction to comparative aspects of limbs, let us quickly rehearse the comparative patterns evidenced by the trunk.

5.2.2 The Comparative Pattern as Evident in the Trunk: a Summary

In the body-plan for very primitive vertebrates, the dorsal (sensory) and ventral (motor) roots of the spinal cord are separate. The ventral roots are in line with one set of 'somitomeres' that give rise to the muscular structures. The sensory dorsal roots supply the structures arising from the next 'somitomere'. These two sets of 'somitomeres' alternate along the length of the trunk. The two together are the equivalent of the 'somite' that exists in other species. This arrangement is typical of both embryos and adults of creatures such as lampreys, and of embryos alone of sharks and rays. This ancient pattern of somitomeres is even present in humans but only very transiently in the very early trunk. They almost immediately coalesce in pairs to form half the number of somites typical of most vertebrates.

This ancient pattern, however, is still present to a limited degree in the heads of all vertebrates where the cranial equivalent of somitomeres (rhombomeres, for example) is still present, and where some cranial motor nerves (e.g. the nerves of the eye muscles, the 3rd, 4th and 6th cranial nerves) are largely separated from their sensory equivalents (e.g. the sensory nerves of the head, the three divisions of the sensory parts of the 5th cranial nerve).

However, in most animals today (adult sharks and rays, and both embryos and adults of fishes, amphibians, reptiles, birds and mammals) the body-plan of the trunk has become changed so that the sensory and motor roots join in relation to the junction of somitomeres into somites. This produces a single spinal nerve for each segment that is "mixed"; it receives sensory fibers and provides motor fibers to a complete embryological segment of the trunk.

The spinal nerves branch into dorsal (dorsal primary ramus) and ventral (ventral primary ramus) nerves. These mixed dorsal and ventral primary rami should not be confused with the dorsal sensory and ventral motor roots emanating from the spinal cord. In fishes, the dorsal ramus supplies the dorsal skin and muscles, and the ventral ramus the ventral skin and muscles. There is a distinct division between dorsal and ventral at what is called the lateral line of fishes. In comparative anatomical terms, then, the lateral line of fishes corresponds to the embryological position of the division of the myotome into dorsal and ventral moieties (as described above) and is very ancient. In vertebrates other than fishes, this separation is less clear.

5.2.3 The Trunk Pattern Ordered Over into Limbs

Multisegmental Formation and Cranio-caudal Disposition

The distribution of the dorsal primary ramus is unrelated to the production of limbs. The ventral primary ramus alone supplies the limb appendages (the lateral fins in fishes and related aquatic creatures, see later, and the limbs in land vertebrates). The appendages are solely ventral structures existing, in fishes, below the lateral line, but often in other vertebrates looking as though they were attached dorsally; this latter is only an impression, they are ventral structures.

The number of segments contributing to the limbs across the vertebrates is extremely variable, as many as 55 in some fish, as few as 3 in some amphibians and reptiles. In humans, as we have seen, there are generally 5 to the limb muscles and an extended number, 9 or 10, to limb skin. As the ventral primary ramus of each segment enters the limb it combines with the others to form a nerve plexus (the brachial plexus in the upper limb, the lumbo-sacral plexus in the lower limb). In other words, the individual ventral primary rami that contribute to these plexuses are arranged in a cranio-caudal series, as are the various muscles and skin areas supplied by them.

Dorso-ventral Patterning

In comparing animals from species to species it is evident that the muscular blocks that we have already recognized developmentally are organized (irrespective of their particular variations) as comparative dorsal and ventral muscle sheets. In the comparison of most species, the developmental picture in the limb is maintained. The ventral primary ramus of each segmental spinal nerve is divided into secondary dorsal and ventral divisions and supplies, respectively, these dorsal and ventral muscles sheets within the limb. Thus, the ventral primary ramus of the spinal nerve, when it is at a segmental level where it participates in a limb, gives rise to a secondary dorsal division and supplies those muscles (dorsal in position) that move the appendage in a dorsal direction (often called extension). Likewise, there is a secondary ventral division that supplies the ventral muscles that move the appendage ventrally (often called flexion).

In fish, these elaborations relate to the pectoral and pelvic fins. The immediate functions of the muscles in the fins are to extend (move dorsal-wards) and flex (move ventral-wards) the fins. However, the muscles on the preaxial border of the fin (both the dorsal and ventral components) move the fin forwards, and those on the post-axial border move it backwards. Further, in many species the fin is also tilted so that the preaxial (cranial) border is both cranial and dorsal, and the postaxial (caudal) border is both caudal and ventral. As a result, the most important movements of the fins (for control in swimming) involve various combinations of these individual muscle groups that, acting in a controlled sequence, produce rotational movements of the fins not unlike a 'figure of eight'.

This rotated arrangement in fish is reminiscent of some of the rotated positions and movements of the limbs found in land vertebrates. It is unlikely, however, that this rotated situation in fish fins has any direct relationship to the various rotations that exist in land vertebrates and therefore humans. Does the terrestrial situation represent a continuation of a fundamental vertebrate arrangement in some fishes? Or is it a secondary rearrangement (as seems more likely), derived through parallel evolution in land creatures? We do not really know. However, the same dorsal and ventral, and preaxial and postaxial arrangements, have certainly been continued into land animals and therefore into humans, and are associated with the yet more complex movements permitted in the limbs of these species.

5.2.4 Patterns Specific to Limbs

Superficial/Deep Divisions and Proximo-distal Migrations

This level of organization, already clearly evident in development, as we have seen, is also seen in animal comparisons. Though not obvious in fish, the dorsal and ventral muscle blocks in amphibians, reptiles, birds and mammals are, each, subdivided into superficial and deep sheets of muscle (compare development, Figs. 4.8–4.11, earlier). This is the same pattern evident in humans.

The superficial dorsal (extensor) musculature is always clearly recognizable in the various animal groups. At the shoulder, one component of this is a superficial muscle block that passes from the dorsal aspects of the scapula to the humerus (the teres major muscle, and its much larger relative the latissimus muscle block, the origin of which has migrated proximally and dorsally as far as the midline of the trunk). At the hip the superficial muscle block passes from the dorsal element of the pelvis (ileum) to the femur (gluteus maximus).

The deep dorsal (extensor) musculature is also always clearly recognizable in the various animal groups. At the shoulder the deep musculature is in two parts. One passes from the trunk to the girdle: the serratus muscle block (comprising levator scapulae and a complex serratus muscle). The second passes from the girdle to the humerus and includes some scapular muscles (e.g. subscapularis) from the dorsal bony element, the scapula, inserting close around the humeral head. (This group has nothing to do with the dorsal scapular muscles of mammals that are, paradoxically, ventral muscles – see below).

At the hip, the main muscles concerned are the lesser glutei passing, again, from the dorsal element of the pelvis (the ileum) to the femur.

The superficial ventral (flexor) musculature is also clearly recognizable. At the shoulder this is present as the superficial pectoralis sheet (pectoralis major in humans) that has migrated proximally and ventrally onto the trunk; as a result it originates from the sternum and ribs and passes to the humerus.

The deep ventral muscle block is represented doubly, first by deep muscles that pass from more laterally on upper ribs to the coracoid of the scapula (lesser pectoralis muscles, pectoralis minor in humans) and second by deep muscles passing from the ventral components of the girdle (the coracoid) to the humerus (the various coraco-humeralis muscles, coracobrachialis in humans). The coraco-humeralis block is very complex in many species with many muscular slips passing from the coracoid to the humerus and even more distally into the forearm. But in humans it is very much simplified, with only a single coracobrachialis muscle attaching at the points: coracoid process and humerus, as indicated by its name.

The ventral superficial (flexor) musculature is also clearly recognizable at the hip. This muscle block is present as the superficial adductors and hamstring muscles that arise from the ventral aspects of the pelvis (the pubic and ischial rami). These origins and these muscles are very extensive. As a result of the changed position that occurs in the lower limb during development, they actually end up lying with their origins in different position in the different creatures. This provides them with different functions.

Thus the adductors, coming from the cranial ventral element of the pelvis (the pubis) to the femur have become protractors of the hindlimb in quadrupedal creatures, but adductors in bipedal humans. Likewise, the hamstrings, coming from the caudal ventral element of the pelvis, the ischium, are thus hip retractors in quadrupedal animals. But because, in humans, the ischium has moved during a developmental rotation into a dorsal position, they have become hip extensors.

The deep ventral muscle block is represented doubly, first by deep muscles that pass from more laterally on upper ribs to the coracoid of the scapula (lesser pectoral muscles, pectoralis minor in humans), and second by deep muscles passing from the ventral components of the girdle (the coracoid) to the humerus (the various coraco-humeralis muscles, coracobrachialis in humans). The coraco-humeralis block is very complex in other species with many muscular slips passing from the coracoid to the humerus and even more distally into the forearm. But in humans, it is very much simplified, with only a single coracobrachialis muscle attaching at points, coracoid process and humerus, as indicated by its name.

The ventral deep muscle layer at the hip is readily identified and unexceptional. It comprises the obturator muscles from the ventral parts of the pelvis to close around the femoral head.

Some Scapular Muscles – a caveat

As mentioned above, there is one group of muscles at the shoulder that, in most mammals (and therefore in humans), seems to occupy a paradoxical position. This group comprises the supra- and infra-spinatus and teres minor muscles. These pass from a dorsal element (the blade of the scapula) to a ventral element, the greater tuberosity of the humerus, hence a paradox. However, in many other forms, these muscles actually originate from the ventral coracoid. In other words, their dorsal origin in mammals is because, in the course of evolution, their origins have migrated dorsally from the ventral coracoid onto the dorsal scapular blade. They are really ventral derivatives in many other species.

As with the developmental bases of limb structures, so these comparative considerations are also fundamental in understanding why structures are the way they are in humans.

5.2.5 The Implications of Function

As with the comparative anatomical relationships of structures, so too, it is possible to recognize underlying principles of functional adaptations of structures. Throughout both the developmental and comparative discussions, function has been consistently mentioned. Thus, further consideration of function is useful at this point. Though the various vertebrates are all very different, certain commonalities of function exist. Common biomechanical principles are at work, though we must take great care to recognize that biomechanics on land are completely different from biomechanics in water, and somewhat different again (though less so) than biomechanics in the air. Thus 'biomechanical optimization' of structures resulting from evolutionary processes and 'epigenetic adaptation' involving reverse engineering during ontogeny produces functional changes in structures that are superimposed upon the developmental blueprints and comparative body-plans previously described.

'Limbs' in Water

Though the primary locomotor movements in fishes are carried out by lateral bendings (waves) that pass along the trunk, further refinements of movements are produced by the 'limbs', that is, fins. The movements of the fins in fishes and other similar forms include movements dorsally and ventrally, hence carried out by the corresponding dorsal and ventral muscle blocks. The terms: extension and flexion, and depression and elevation, are also used for these same movements. By employing combinations of fore and aft, and dorsal and ventral muscle slips, the angle of attack at which the fin meets the water can be changed and, at its most extreme, the figure-of-eight movements described previously are possible. These complex movements help the fish to limit, control, or produce: yawing, pitching and rolling movements. They are superimposed upon the primary locomotor movements that drive the creatures forwards (lateral waves of the trunk produced by the axial, trunk, musculature). Similar complex movements occur in limbs that operate 'in air', that is, wings in birds.

Limbs on Land

Case 1

In many land vertebrates, for example, amphibians and reptiles, the principal movements of locomotion are, like those in fish, still produced by lateral waves of the trunk musculature. However, these waves move the trunk by using limb contacts with the ground as the fulcra for movement, rather than direct contact of the trunk skin with the water (as in an aquatic medium).

In such creatures, movements of the limbs, which are, in many of them, held in a somewhat sprawled position, laterally from the trunk, also contribute to locomotion. These movements require not only the dorsal-ward movement (lifting the foot from the ground as produced by the dorsal muscles) and the ventral-ward movement (pressing the foot against the ground as produced by the ventral muscles), but also movement of the raised limb forward into position for the next ground contact (limb protraction) and backward movement of the lowered limb fixed on the ground as an addition to driving the body forward (limb retraction).

Thus protraction attempts to draw the limb cranially, and because, during this phase of locomotion the limb has been lifted from the ground, the movement actually swings the limb cranially. For that particular limb at that point this is called the 'swing' phase. This is the limb portion of the return stroke of locomotion and is produced by both dorsal and ventral muscles that lie cranial to the axis of the joint.

In contrast, retraction attempts to draw the limb caudally but because, at the same time, the limb is being pressed against the ground, the body is thrust forward over that limb, that is therefore called the stance limb. This portion of the power stroke is produced by those dorsal and ventral muscle blocks that are behind the joint axes. They provide additional impetus to the forward locomotion of the animals produced by the lateral sinuosity of the trunk.

Case 2

In some reptiles and especially in mammals there are other differences in limb function in locomotion. Thus in these creatures the elbows point more caudally and the knees more cranially (rather than both being positioned somewhat more laterally as in Case 1). In addition, these limbs are positioned more beneath the body than to the side, though this varies in different species (of both mammals and non-mammals). The trunk contribution to locomotion in these creatures still involves a degree (in humans only a small degree) of a lateral wave (lateral flexion); of greater importance is a dorso-ventrally directed wave (flexion and extension) of the trunk.

The limb movements, retraction (the power stroke, the stance phase) and protraction (the return stroke, the swing phase), together with the dorso-ventral movements of the trunk, are a correspondingly much greater portion of locomotion. Combinations of both dorsal and ventral muscles produce these locomotor movements of the limbs and trunk. Thus caudal muscles (of both dorsal and ventral muscle blocks) that would move the free limb caudal-wards combine in their actions when the limb is on the ground, forcing the trunk forwards, the power stroke of locomotion.

The cranial muscles (of both dorsal and ventral muscle blocks) combine in their actions when the limb is off the ground to swing the limb forwards, the return stroke of locomotion. The dorsal/ventral flexions of the trunk provide increased 'step length' for these limb movements.

Other factors are also involved in this type of quadrupedal movement. One of these is the movement of the very large heavy liver (attached to the diaphragm) that swings cranially and caudally on the diaphragm during the steps. It therefore acts as a sort of piston that increases the trunk contribution to locomotion. Further, it especially acts on the movements of respiration; the forwards and backward movements of the liver and diaphragm driving coordinated inspiration and expiration of the lungs. These mechanisms are especially important when very fast movement in large animals is required; it been especially studied in race horses.

Case 3
In humans many of these movements are rather similar to those just described for mammals in general. They show, however, further differences that relate to the general arboreal environment within which human precursors evolved. That is, the more recent upright position of the body in bipedality, and the consequent partition of function of the limbs into non-weight-bearing upper limbs (though still involved in non-weight-bearing ways in locomotion) and weight-bearing lower limbs. There are three sets of comparisons.

The first is the increased flexibility of limbs, particularly of upper limbs (especially shoulders) and to some considerable degree also of lower limbs (especially hips), that can cope not only with the functional demands of the three-dimensional arboreal environment of human ancestors, but are also useful for many non-locomotor activities of humans today.

The second includes relative reductions in muscle and bone masses for the functional demands of the non-weight-bearing human upper limb relative to the lower limb, compared with more equal balance in many non-human primates using all four limbs for locomotion.

The third involves modifications of hands and feet such that the large powerful grasping hands of non-human primates that climb have become, in humans today, smaller less powerful hands that are nevertheless capable of a very wide array of activities in non-weight-bearing modes for handling the external environment. Likewise, the large powerfully grasping feet, again in animals that are still capable of climbing, have in humans remained large but become non-grasping solid feet taking the entire body weight during movement on two limbs.

5.2.6 Limb Rotations, Torsions and Twists

The various modifications of limbs implied by these developmental and evolutionary discussions have resulted in limb structures being twisted. These twists are all-pervasive in human anatomy, being seen in the relationships of skin, muscles and nerves throughout both limbs.

Thus, we have already looked at the 'rotation' of limbs that occurs in mammals during the 'ontogenetic time' of development from limb buds to actual limbs. There are, in addition, also torsions of limbs that result from the 'deep time' of evolution from fins to limbs, as described above. It has sometimes been supposed that these two reflect one another. In fact, they are actually separate (even though, at some deep complex level, there may possibly be a relationship between them). Finally, there is a third 'twist' of the upper limb that is found in primates, especially humans. This last is due to the way human anatomists initially described the standard position of the human in relation to our bipedal stance. It has been carried over into such non-human species as apes and monkeys (that can also supinate their fore-arms), but not into such non-primate species as dogs and sheep (in which the forelimb is fixed in the pronated position). There is room for confusion here and we will attempt to sort this out.

Thus the human limb, being at early stages a laterally directed outpouching of the trunk, has dorsal and ventral surfaces, cranial and caudal borders, and cheiridia with developing thumbs and big toes on the cranial borders, and little fingers and little toes on the caudal borders. As development proceeds, the limbs come to be rotated. Thus in the fetus and proceeding thence into the baby, in the upper limb the arm is adducted and extended (compared with a general quadrupedal mammal) so that the elbow comes to lie against the trunk. As a result, the elbow points dorsally. In the lower limb, the thigh is adducted and flexed (so that the thigh also comes to lie in line with the trunk). But the knee points ventrally.

In addition to these changes in position of the arm and thigh, there are equivalent medial rotations of the forearm and leg. They produce positions of the hands and feet, so that the originally cranially pointing thumb and big toe both become medially directed. Likewise, the caudally pointing little fingers and little toes become laterally directed. These are also the positions that these parts occupy in the adults of most quadrupedal mammals.

If one then takes the folded hindlimb of the fetus, or even baby, and places them in the bipedal position (or watches them grow through infancy into the bipedal position), then the folded limbs open up, the lower limb coming to lie parallel to the body axis, the knee pointing forwards, and the big toe placed medially. If, however, one does the same thing with the upper limb, then, because of the original starting positions, the limb again comes to lie alongside the body axis. The elbow, however, now points backwards and the thumb points medially. Both of these manipulations also replicate the state in quadrupedal mammals save that the limbs are in line with the trunk rather than being at right angles to the trunk.

5.2.7 Limb Rotations: a *Caveat*

Unfortunately, medical anatomy has chosen to define the standard position of the upper limb as with the palms facing forward. (This is for a fairly good reason – that the palm is truly a ventral aspect of the hand). However, this convention is not carried forward into the foot where the sole is truly the ventral aspect of the foot. This is also for a very good reason; because the leg cannot be pronated and supinated as the forearm, so the foot cannot be placed with the sole looking forward. This different position of the forearm and hand mean that, in the medical definition, the thumb is directed laterally. This is achieved by placing the forearm in extreme supination. The result of this is that, whereas in the lower limb the whole process is a medial torsion of the entire limb, in the upper limb, the arm is also twisted medially, but the forearm comes to be twisted laterally (Figs. 5.6–5.8).

These twists in both limbs and this difference in the upper limb alone can be readily seen from two views of the anatomy of the limbs. One of these examines the form of the entire limb with consideration of how the skin areas and the muscle blocks have become twisted (Figs., 5.6–5.8). The other examines the positions of the internal structures of the entire limb through changes in positions of nerves in cross-sections (Fig. 5.9(a–c)).

Though it may seem curious that we should be looking at these developmental changes in a section on comparison and function, the rationale for understanding becomes clear as we examine different limb functions.

Let us look first at the main proximal segment of the limbs: arm and thigh, respectively. For example, the twist in the arm, with the elbow pointing backwards, means that the dorsal skin comes to lie at the back of the arm. The dorsal muscles (e.g. triceps) likewise come to lie at the back. The dorsal nerves (e.g. the radial nerve) likewise lie at the back. These **dorsal** positions fit with the medical terminology of **posterior** in the arm. The twist in the lower limb, with the knee pointing forwards, means that the dorsal skin and the dorsal muscles (e.g. the quadriceps femoris) and the dorsal nerves (e.g. the femoral nerve) come to lie on the front of the thigh. The positions of these **dorsal** structures do not fit with the medical terminology of **anterior** in the thigh.

Similar considerations come to pass in the muscles and nerves of the limbs. In the lower limb the embryologically **dorsal** surfaces, muscles and nerves of the leg and foot come to be in the medically designated **anterior** positions (if by that we mean pointing forward). The embryologically **ventral** surfaces, muscles and nerves of the leg and foot (sole of the foot) are lying, in medical terminology, in **posterior** positions (if by that we mean pointing backwards). However, because of the medical definition of the 'anatomical position' the opposite is the case in the upper limb. The supinated position of the forearm has returned the embryological **ventral** structures: surfaces, muscles and nerves, to the medically defined **anterior** position; likewise, the embryologically dorsal surfaces, muscles and nerves have been returned to the **posterior** aspect. This is important both for truly understanding the science behind the anatomical and functional patterns, and for the implications in clinical medicine.

As indicated, these positional problems relate to muscle function as well as skin areas. As a result, there are distinct functional implications. Thus the embryologically dorsal muscles (embryologically extensors) of the thigh, leg and foot all produce extension of the limb. This is the position that the limb takes up during, for instance,

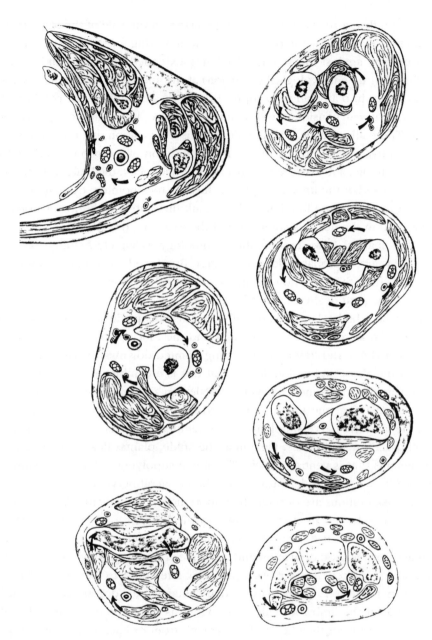

FIGURE 5.9

Rotations as seen from transverse sections of the upper limb: (a) transverse section at shoulder, (b) transverse section at shoulder and down the arm with structures exploded and with arrows showing rotations of nerves around the arm axis, and (c) transverse section in forearm with structures exploded and with arrows showing reverse rotations of nerves around the forearm axis.

the placing reflex, a reflex stimulated by pressing on the sole of the foot, as by the ground in standing, or by the physician's hand in testing for the placing reflex. The limb is pressed against the stimulus by being extended at the hip and knee, and 'dorsi-flexed' (true embryological extension) at the ankle and foot joints. As a result the limb moves into a position to support the body, hence the term 'placing reflex'.

In contrast we can consider the situation of the lower limb embryologically ventral muscles. These all produce limb flexion. This is the position that the limb takes up in a nociceptive reflex, a reflex due to the application of a noxious stimulus to the foot – for example, a nip on the toe. In a nociceptive reflex, the limb becomes flexed at the hip, flexed at the knee and 'plantar flexed' (toes pointed, true embryological flexion) at the ankle and foot joints. As a result the limb is withdrawn (protectively as it were) from the noxious stimulus (but the toes are pointed).

Does this matter? If we look at the terminology as merely helping to describe where things are, it probably does not matter. But if we look at understanding why things are where they are, how things work, even how they function and react, it matters a very great deal (see also Chapter 3).

For example, a ballet dancer (especially), but also people in general, think of lower limb extension as the limb being straight and the foot pointed. That is: thigh and knee stretched out and foot and toes being pointed downwards. However, consideration of the above discussion shows that this is a combination of embryological extension at the knee and embryological flexion of the ankle and foot, plantar flexion. Extension at the knee is produced by quadriceps femoris. This is a dorsal muscle, and the nerve supplying it, the femoral nerve, is a dorsal derivative of the lumbosacral nerve plexus. Flexion at the ankle, plantar flexion, is produced by gastrocnemius that is ventral muscle. The nerve supplying it, the tibial component of the sciatic, is a ventral derivative of the lumbosacral nerve plexus.

These considerations are of scientific importance in many fields. A surgeon transplanting a muscle and its nerve to take up the function of some other damaged muscle and nerve is best advised to use, as the replacement muscle, one that has the same embryological derivation and the same embryological nerve supply, as the muscle being replaced (if possible). A physician attempting to understand the effects of specific nerve lesions needs to know that damage to the dorsal nerves of the limb nerve plexus at the limb root will cause paralysis of all the dorsal muscles of the limb – whether or not they lie on the 'layman's front or back of the limb. A pediatrician, seeing Klumpke's paralysis (flexed and medially rotated shoulder, extended elbow, pronated forearm and flexed hand and fingers) needs to understand this discussion. This is also so for understanding the baby's grasp reflex (shoulder, elbows, hand and fingers flexed). An anthropologist looking at a variety of related creatures needs to know how these embryological differences in limb structure relate to function, development and evolution.

5.2.8   The Limb Nervous System

The peripheral nervous system is treated several times in this book in order to demonstrate its integration with the various parts: trunk, head, and central nervous system. It is useful here to summarize that part of the nervous system that relates to limb organization.

We already know that the structures in each segment (somite) are supplied by a spinal nerve formed by the union of dorsal (sensory) and ventral (motor) roots of the spinal cord. This spinal nerve then almost immediately undergoes a fundamental division into a dorsal primary ramus and a ventral primary ramus. These rami, therefore, are all 'mixed' nerves, that is, they each contain fibers from the dorsal (sensory) and the ventral (motor) roots. This division into dorsal and ventral rami corresponds to the division of the somite of the trunk into its dorsal (nervous system influenced) and ventral (notochord influenced) parts. In the trunk, the dorsal primary ramus of each spinal nerve supplies the dorsal derivatives of the myotome, dermatome and sclerotome. In the adult, therefore, it supplies the dorsal spinal extensor muscles, the dorsal skin close to the midline on the back, and the dorsal bone (part of the transverse processes, neural arches and spines) to which these muscles are attached and which surround the nervous system. It does not supply any part of the limb because the limb is a ventral derivative of the trunk.

The (much larger) ventral primary ramus of each spinal nerve supplies the more extensive ventral derivatives of myotome, dermatome and sclerotome. In the adult, each therefore supplies the much larger group of muscles of the dorso-lateral, lateral and ventral body wall, the much larger skin area dorso-lateral, lateral and ventral on the trunk, and the ventral skeletal elements (vertebral bodies, ventral parts of the transverse processes, costal processes or ribs, and the sternum of the trunk).

However, these ventral primary rami give, in the trunk regions where limbs appear, branches into the limbs. These branches divide into secondary dorsal and ventral branches corresponding to the (secondary) dorsal and ventral divisions of the limbs. They supply the skin areas of the limb derived from the ventral trunk skin areas, the muscles of the ventral parts of the somites that have migrated into the limb from the corresponding trunk segments, and the connective tissues (bones, joints, fascial sheets, connective tissues in general, and vascular channels, arteries, veins, and lymphatics) that have arisen from that part of the lateral plate mesoderm that has streamed from the ventral aspect of the trunk into the entire limb.

In passing from the anterior primary rami to the named individual nerves of the limb, a new structure is formed. This is called a nerve plexus. Its effect is to bring the various anterior primary rami together at the root of a limb in which trunk segmentation seems to be hidden. This allows all the nerve fibers from the many segments to pass into the limb through a fairly narrow portal. Yet the specific structures of the plexuses maintain the segmental arrangement so that correct segmental distributions are made more peripherally in the limb. These plexuses are, for the upper limb, the axillary plexus, and for the lower limb the lumbo-sacral plexus.

5.3   Limb Variations

Although anatomical texts show mostly the 'average' human, in fact there is a range of variation in limb structures. Though evident in all structures throughout a limb: skin, muscle, bone, joints and fascial sheets, variation is often most evident in muscles, perhaps because muscles are so well known from centuries of dissection. Some of these variations relate to the mechanical functions of the parts; some are genetic in origin; some are the result of developmental processes. Of course, many stem from combinations of these different causal factors.

5.3.1   Functional Variations

Thus, knowledge of the sizes, structures and attachments of muscles can, through anatomical inference, provide some understanding of muscle function, of how muscles are used in posture and movement. Such inferences, the main source of evidence about function in earlier times, can now be tested using methods like electromyography, cinematography, cineradiography, mechanics, kinetics, and so on.

Further, the loads on bones that stem from muscle function give information about the mechanical efficiency of bones, that is, the degree to which contraction of muscles procures relief of stress on bone. These particular functions, mechanical stress relief, are often neglected, but can now be tested using the methods of strength of materials, both theoretical and experimental stress analysis.

Thus muscles (and their associated tendons, ligaments and fascial sheets) illuminate how tension bearing affects the mechanics of hard tissues (bones and cartilage). This is known as **tensegrity** (a term invented by Buckminster Fuller: compressive int**egrity** through **ten**sion bearing). That is, though much compression in a part is borne by the bony struts in that part, a considerable portion of **compression** is actually relieved through **tension** developed in appropriately aligned connective ties: fasciae, ligaments, and muscles with their tendons and aponeuroses (flat tendinous sheets). At first glance this sounds paradoxical but it can be tested by mechanical modelling studies. All such tests aid in understanding the nature of functional adaptation of anatomical structures.

However, in addition to such functional tests a large number of muscular variations in humans imply function by inference rather than testing. Many of these are catalogued in the anatomical journals of the last three centuries. That a very large number have been recorded for humans is because of the large number of human cadavers that have been dissected in the medical schools of the past. Such knowledge has always been important in clinical medicine, especially for surgical procedures. Today, when surgical procedures are so much more complex, so much more frequently guided by non-invasive imaging, and so often done 'scopically', information about such anatomical variations has become ever more important.

Some muscular variations imply function through statistical comparisons with closely related species. Thus, though the caudal limit of the origin of pectoralis major on the trunk in humans is usually the 6th rib, in some humans it can extend caudally to the 8th rib. (I once had a student who was so angry that **his** body was not the 'same as the book'. I think he rather wanted to exchange it for one that was 'like the book'). This human range contrasts with the extent of the origin in many Old World monkeys in which it can be as caudal as the 9th rib on an already much cranio-caudally elongated thorax. In apes the origin may be extend only as caudally as the 5th rib in a much shorter barrel-shaped thorax.

Such differences are likely of functional import in relationship to differences in locomotion and posture. Thus, in monkeys, because the large muscle extends far caudally, on an already very long and narrow thorax, it is well suited to producing the upper limb retraction that is such an important part of the power stroke below the trunk in quadrupedalism (whether on the ground or in the trees). In apes, where the very large muscle is placed much more cranially in terms of rib count, and even more cranially because the thorax is so short, it is thus able to function more efficiently within a raised arm in front of or above the head in suspension of the body during many climbing and acrobatic movements in the trees. Yet at the same time that powerful muscle is still able to assist in retraction in the quadrupedal movements of which apes are also capable when knuckle-walking or fist-walking on the ground. In contrast, in humans the muscle is relatively much smaller and is in an intermediate position on the chest wall. It is not involved in any very powerful weight-bearing locomotor movements, though it clearly can participate in a very wide range of movements, for example, during communication by semaphore, activity on the trapeze, or in throwing the javelin.

Many upper and lower limb muscles in humans and related primates show equivalent statistical variations that make sense in terms of differences in function and lifestyle.

5.3.2 Genetic Variations (?)

There are other types of anatomical variations for which a functional implication seems unlikely. Thus, some statistics of muscular variation seem to imply that variations can be related to differences between closely related species that are also very similar functionally. For example, in the arm, the deep head of the coracobrachialis muscle, usually absent in humans, was present in 10 of 14 specimens of night monkeys (*Aotus*) and in all 6 specimens of titis (*Callicebus*). In contrast, this muscle was usually absent (8 out of 10 each) in uakaris (*Cacajao*) and sakis (*Pithecia*). Though all four of these creatures move rather similarly, they are thought to be in two different higher taxonomic groupings (the sub-families: Aotinae and Callicebinae, respectively). Is it possible that it is the taxonomic difference (perhaps of genetic causality) to which this is due?

This is only one of a large number of such differences. For example, in the thigh, the gracilis muscle had only a single belly in all 14 specimens of night monkeys and titis, but there were two bellies in all 10 specimens of uakaris and sakis. These similarities in night monkeys and titis, and their differences between uakaris and sakis, are unlikely to be of much functional importance. It is more likely that it is because these two respective pairs of New World monkeys are in different subfamilies.

In humans there are a number of such similarities especially in identical twins, in individuals within families, even with local close-knit communities. Many of these seem not to be particularly functional but could be interpreted as characters giving information about individual and population differences, possibly therefore, about genetic relationships.

5.3.3 Developmental Variations

A further type of muscular curiosity seems related to development. An example is the deltoid muscle. In many land vertebrates: amphibia, reptiles and non-placental mammals, the deltoid consists of two parts: a ventrally located deltoideus acromio-clavicularis (originating from the ventral components of the girdle, the clavicle and the scapular acromion) and dorsally located spinodeltoideus (stemming from a dorsal part of the girdle, the scapular spine). These heads are each supplied by their own branch of the axillary nerve. In most primates, including of course humans, there are, in contrast, three parts to the deltoid: ventral (anterior in human medical terminology), middle and dorsal (posterior in human medical terminology). Yet these three heads in primates are only supplied by two branches of the nerve. One branch passes to the dorsal deltoid and the other to both middle and ventral deltoid. Why should the pattern of innervation (two branches) differ from the muscular arrangement (three heads)? The obvious possibility is that in primates, in contrast to reptiles, one of the heads of deltoid has split giving three heads.

However, examination of macaque (*Macaca mulatta*) fetuses supplies a possibly different answer. In the macaque, deltoid first appears as a single muscle precursor that later becomes partially separated into three parts. The developing axillary nerve supplying deltoid, likewise, has three branches that seem to be directed towards these three parts. In slightly older fetuses one part of deltoid disappears, together with the branch of the nerve reaching towards it. Such losses are common in development: many features in the earlier stages of embryos and fetuses disappear later. What next happens, however, is that, at a slightly later stage still, one of the remaining parts of the muscle (the ventral-most remnant) undergoes a secondary split, thus giving rise to new ventral and middle heads. This returns the number of heads to three. One of the two remaining nerve branches then comes to innervate both new muscular heads. It is not common to examine later stages of development; unless this is done, such changes, and the science behind them, may be missed.

5.3.4 Specific Example of a Variation: Dorsoepitrochlearis

Some examples of muscular variations seem to relate to comparisons with other primates. Thus, a dorsoepitrochlearis muscle is present in all non-human primates. This is a muscle arising from the tendon of latissimus dorsi and passing into the upper limb to be variously inserted more distally.

In quadrupedal monkeys it inserts into various structures (mostly fascial) in the arm and forearm and is thus capable of extending the elbow (retracting the forearm) as well as contributing to the function of latissimus dorsi in extending the shoulder (retracting the arm) at the same time. Both of these movements are part of the power stroke, retraction, of quadrupedal locomotion.

However, in apes and prehensile-tailed monkeys, the distal fascial attachments of dorsoepitrochlearis do not extend below the elbow and thus the muscle assists extension (retraction) at the shoulder without impeding flexion at the elbow. This is a combination of movements more useful in the under-branch hanging, climbing and swinging that these creatures utilize.

We might think it curious that this muscle is absent in most humans.

It is present, however, as a variation, in about 5% of human cadavers. Though clearly the equivalent of the dorsoepitrochlearis of other primates, in humans it has received many other names related to other variations in its structure and attachments (e.g. latissimo-condyloideus, accessorium tricipitalis) and may be linked with other variant muscles (e.g. costo-epitrochlearis and chondro-epitrochlearis). When present in humans, the dorsoepitrochlearis variant has the same form that is found in apes and prehensile-tailed monkeys (Fig. 5.10(a, b)). (This is one of the many pieces of evidence implying a climbing ancestry for pre-humans at some point).

Cladistic thinking might suggest that the arrangement of dorsoepitrochlearis, when present in humans, should be interpreted as a human sharing a primitive character with apes. Functional anatomy seems to imply that it is present because humans arose from creatures that had similar arboreal adaptations as apes (and prehensile-tailed monkeys). Both could be true. This is an old story well documented in the literature. It is as though the muscle, when it is present, is a 'ghost' of a prior condition, an 'atavism' in the older literature.

Yet the fascial sheet surrounding the anomalous muscle is present in all humans, whether with or without a muscle, as connective tissue strands that pass from the tendon of latissimus dorsi into the intermuscular septum of the arm. Is this connective tissue element also a 'ghost' indicating, in this case, that the muscle used to be present in humans, is no longer present in most humans, but its fascial components are still present in all humans? This reminds us that the muscular elements derive from myotome components of paraxial mesoderm that have migrated into the arm. The fascial elements, in contrast, derive from lateral plate mesoderm in the arm. If a muscle is absent because of a change in the myotomal structures relating perhaps to hox genes, its connective tissue elements (fascial sheets, tendons), being derived from different developmental elements (e.g. connective tissue factors), may not necessarily be similarly affected.

For the moment, this is just a possibility, though it does stimulate new questions. When muscles disappear over evolutionary time is this sometimes because muscle developmental factors are modified? And could we tell? When rare variations persist is this because a modified muscle developmental factor is still minimally active, a question for molecular developmental biologists? Further, when fascial components and coverings of muscles persist even when the muscles have disappeared, is this

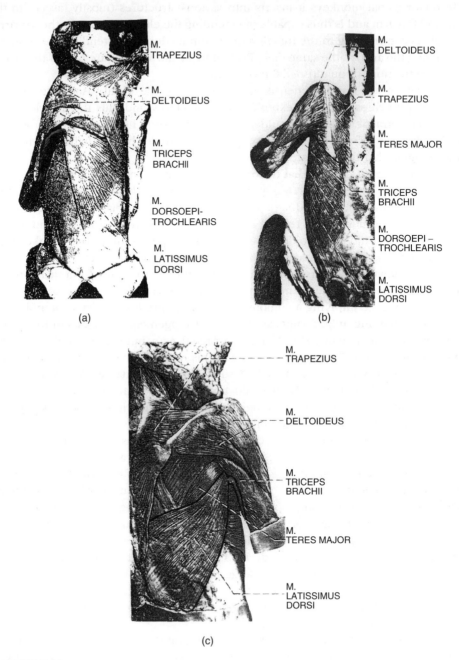

(a)

(b)

(c)

FIGURE 5.10

Examples of the dorsoepitrochlearis muscle in a gibbon (a) and the unusual variant in a human (b).

because connective tissue structures are produced by different molecular factors? In other words, remnants of muscles that have otherwise been lost, and the presence of their related fascial sheets without muscle, may be useful 'double ghosts of the past' that illuminate adult human anatomy at the interface of development and evolution.

5.3.5 More on Muscular Variations: Scapular Muscles

Though it might seem that the days are gone when new information could be revealed by dissection, in fact, the above examples show that this is not so. Of course the comparative anatomies of bones, muscles, nerves, and blood vessels are well known. They are easy to observe, dissect, and measure. The comparative architectures of tendons, ligaments and aponeuroses are, however, somewhat less well known. The comparative arrangements of connective tissue fascial sheets are scarcely known at all. These various connective tissue structures are so much more difficult to dissect than bones, muscles, tendons, nerves, and blood vessels, particularly in fixed specimens. (It should be pointed out, however, that such structures are very clear and easily observed in fresh or frozen cadavers, and during post-mortem examinations and in surgery.) If the above ideas are correct, the arrangements and variations of fascial sheets may supply new information about the development and evolution of human anatomy.

Another example of muscles around the shoulder where straightforward anatomy seems to belie true relationships are the infra- and supra-spinatus muscles. These seem to be obviously dorsal limb muscles, lying as they do on the dorsal surface of the scapular blade; they are certainly so described in most anatomical texts; they lie even more dorsally than the clearly dorsal subscapularis muscle on the under surface of the scapular blade. This seems to be confirmed because the scapular blade is, in both embryological and evolutionary terms, a dorsal element of the shoulder girdle.

Yet the nerve supply of these muscles suggests that they are actually in all other ways: anatomical, embryological, comparative, and evolutionarily, ventral in origin. Thus the individual bundles of their innervating supra-scapular nerve branches can be dissected proximally as they pass through the various components of the brachial plexus and into the ventral primary rami. Here they lie with other fibers destined for the ventral divisions of the trunks of the brachial plexus.

The evidence of comparative anatomy: the disposition of these equivalent muscles in marsupials and reptiles, also suggests that they are really ventral in origin. In these forms they arise from the coracoid process (a ventral girdle element). In mammals, it seems likely that they have migrated dorsally from the ventral coracoid to the dorsal scapular blade. Connective tissue bands extending from the coracoid to the shoulder capsule (the tendons insert on the joint capsule as well as the greater tuberosity) may be the ghosts that confirm what the nerves imply.

Where else in the human body may muscular losses like dorsoepitrochlearis and muscular rearrangements like those of the dorsal scapular muscles, have occurred? One answer may lie in the upper limb as a whole. In humans, in contradistinction to all other primates, the upper limb is not used in powerful locomotor movements. This implies that reduction, perhaps even loss, of upper limb musculature might not produce major functional deficits. This would not be expected in hindlimb musculature. Is it possible that this is a broad theme?

5.3.6 Ventral Muscles and Fascia in Limbs

As we have already seen, there are several sets of forelimb muscles that derive from individual ventral somite sheets. Some of these are girdle-to-limb muscles. These have serial homologs in the lower limb and their condition can not only be compared between the forelimbs of humans and non-humans, but also in relation to equivalent differences in the hindlimbs of humans and non-humans. Others are trunk-to-girdle muscles (sometimes with extension into the proximal parts of the limbs) and these do not generally have hindlimb serial homologs (or such homologs are very small in creatures where the trunk-to-limb girdle arrangements do not permit mobility in the hindlimb) that hindlimb comparisons are irrelevant. All of them belong either to the ventral or dorsal division of the limb. Let us examine them group by group.

Girdle-to-Limb Cranial Ventral Muscles

We already know that two layers of embryologically ventral muscles acting at the hip arise from the embryologically ventral bony elements of the pelvis: the pubis and ischium. (A reminder: though the ischium is anatomically dorsal in location in the human adult, it commences as a ventral element in the fetus and only later becomes dorsal in position through the twist and re-angulation of the limb that occurs during development.)

Thus the cranial-most ventral muscles are the various thigh adductors deriving from the cranial-most bony element, the pubis. They are mostly one joint muscles acting upon the proximal limb joint (the hip). In all primates there are many such adductors ranging from the shortest, adductor brevis, through adductor longus and the adductor head of adductor magnus to adductor gracilis. These muscles may each have as many as two or even three heads in different species. As a sheet, however, they are all of similar embryological derivation, all part of the same superficial multisegmental sheet, and all have the same nerve supply, a ventral branch of the lumbo-sacral plexus, the obturator nerve.

In primates in general, they are splayed out as a muscular fan passing downwards and laterally from the pubis onto the medial aspect of the femur and acting on the hip joint. The insertion of one muscular belly, adductor gracilis, has actually migrated below the knee and so also acts on the knee joint. There are also many muscular variations in different specimens and species. Most of these are small superficial bellies with long tendons passing distally into leg fascia. They are similarly complex and big in both humans and non-human primates (Fig. 5.11(a–d)).

The equivalent muscular block in the forelimb comprises the coracobrachialis muscle sheet, also primarily uniarticular, and acting upon the serially homologous joint,

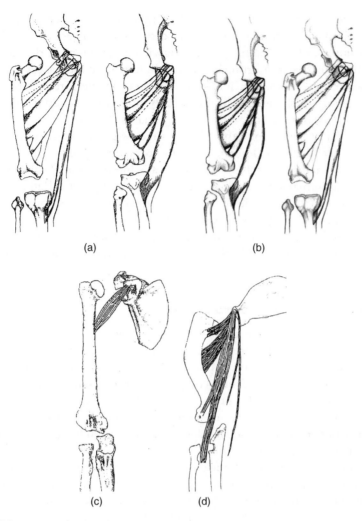

FIGURE 5.11

The cranial ventral muscles of the hip (known as the hip adductor group of muscles) in non-humans (a) in humans (b) and in the equivalent cranial adductor muscles of the shoulder (known as the coracobrachialis group) (c) in non-humans and humans (d).

the shoulder. In non-human primates, varying a little from species to species, this muscular sheet arises (as its serially equivalent component in the lower limb) from the cranial ventral skeletal element, in this case, the tip of the coracoid process of the scapula. The muscular sheet fans out as several muscle bellies. It is extensive, generally comprising a deep head (coracobrachialis profundus) that often has two or more components, a middle head (coracobrachialis intermedius) again with as many as three components, and a superficial head (coracobrachialis superficialis) also often with several components. Each of these passes down the limb inserting into the humerus at correspondingly more distal points. There are, further, several additional muscular bundles that pass into fascia lying even more distally and superficially, with many fibers inserting beyond the elbow joint into fascia in the forearm. In other words, the coracobrachialis muscle sheet in the forelimb of non-human primates is enormously similar to the serially homologous adductor sheet in the hindlimbs in both human and non-human primates (Fig. 5.11(c)).

In contrast, in humans alone this muscle sheet is greatly restricted. There is one component only (called in human anatomy, coracobrachialis, Fig. 5.11(d)) representing only a small part of the coracobrachialis intermedius of other forms. However, individual humans may present, rarely, with a range of variations showing all the other heads of this muscle in non-human primates. When present these remnants are accompanied by their investing deep fascial sheets. Even when they are absent, the fascial sheets remain. On occasion, small muscular bundles (presumably non-functional) can be found in the fascial sheets in the arm, mainly the intermuscular septum.

5.3.7 Girdle-to-Limb Caudal Ventral Muscles

A similar picture exists in the caudal ventral muscles. In the hindlimb these arise from the ischium located more caudally than the pubis (embryologically ventral even though in a dorsal position in humans). These comprise the various hamstring muscles. In contrast to the cranial components, these muscles are primarily two joint muscles passing over the proximal joint (hip) to insert on each side of the distal joint (knee). Though there are some differences among the various primates, the overall arrangement of these muscles is similar in both humans and non-humans (Fig. 5.12 (a–d)).

In the forelimb the equivalent muscle group arises from the root of coracoid and its related upper part of the glenoid (again more caudal than the coracoid tip). This is the biceps/brachialis complex. The bony elements are, again, embryologically, caudal ventral parts of the respective limb girdles. These muscles in non-human species are almost as complex as the hamstring group in the hindlimb (Fig. 5.12 c). They can show as many as three separate biceps muscles inserting into both the radius and the ulna, and up to four partially separate or completely separate brachialis muscles.

In humans, in contrast, these muscles are not only relatively small but reduced in number and complexity, with usually there being only the well recognized single biceps (though with two heads as its name indicates) and single brachialis (Fig. 5.12d). Variations exist in humans, however, in which both biceps and brachialis may be partly or even completely divided, and may present with further additional heads or muscular connections with other muscles just as is the standard case in non-human primates. With all these variations are related investing fascial sheets. Again, then, it would appear that in humans this forelimb muscle group as a whole has become much less complex and much smaller.

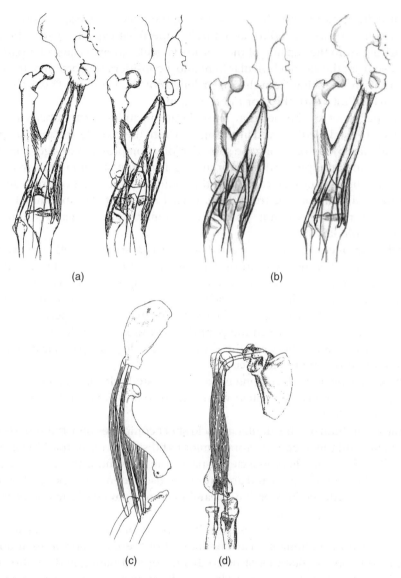

(a) (b)

(c) (d)

FIGURE 5.12

The caudal ventral muscles of the hip (known as the hip hamstring group of muscles in non-humans (a) and in humans (b) and the equivalent caudal muscles of the shoulder (the biceps, brachialis group in non-humans (c) and humans (d).

5.3.8 Trunk-to-Limb Ventral Muscles

The group of ventral shoulder muscles that originate from the trunk and cross the shoulder to be inserted very close to the shoulder joint are the various pectoral muscles. As we already know, these are organized into superficial and deep layers.

The superficial part in non-human primates, the pectoral mass, comprises several different components, including an entirely superficial capsular part (often rather small) arising from the capsule of the sterno-clavicular joint, a clavicular part proper arising also from the sternoclavicular joint and a considerable portion of the clavicle, and a sternal part arising from the major length of the sternum. This latter may be further divided into two, three or more portions.

The deeper layer of the pectoral musculature includes, first, subclavius, arising from the first costal cartilage and inserting into both the clavicle and the clavipectoral fascia (and thereby also crossing the shoulder joint complex). The larger components of this layer include: pectoralis minor arising from the sternum and ribs medially and inserting mainly into the capsule of the shoulder joint, and pectoralis abdominis arising more distally on the trunk from fascial sheets of the abdominal wall and inserting with pectoralis minor. These muscles may also be organized into several clearly separable fascicles.

A final component includes a part that is also associated with the panniculus carnosus muscle. (Panniculus carnosus is primarily a muscle of the superficial fascia and, therefore, technically not a limb muscle, as described elsewhere). Part of panniculus carnosus is, however, attached to a tendinous band lying deep to the pectoralis minor and inserting with the tendon of that muscle. There are frequently a series of connections between the panniculus carnosus muscle and various other shoulder muscles. Some heads of these various muscles may pass far down the arm and even into forearm fascial sheets.

All these conspicuous developments are associated with movements of the limb in creatures (non-human primates) where the forelimb is heavily involved in locomotion (Fig. 5.13(a)).

In contrast, in humans this double muscle sheet comprises only the superficial pectoralis major, and the deeper pectoralis minor and subclavius of traditional anatomical texts (Fig. 5.13(b)). These muscles all pass from the ventral aspects of the thoracic cage, cross the shoulder joint and are variously inserted in the upper end of the arm. They are much reduced in overall mass and in cranio-caudal extent compared to the various non-human primates.

However, all of the extra muscles described above for non-human primates may be found as variations in humans. The commonest of these is a pectoralis abdominalis although additional digitations of pectoralis major and minor, and additional fibers connecting with other shoulder muscles, are all frequent. Even muscles representing the pectoralis-panniculus carnosus interchange can sometimes be found passing from the pectoral muscles to the superficial fascia. Normally in humans there is no panniculus carnosus, though its fascial sheet, the panniculus adiposus, is always present. These variations and persistent fasciae seem to reflect the complex elements of non-human forms.

5.3.9 Dorsal Muscles and Fascia in Limbs

Equivalent comparisons of the dorsal musculature in the hindlimb and forelimb of non-human primates with humans imply that humans have similar restrictions in

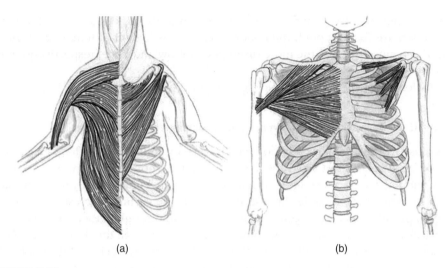

(a)   (b)

FIGURE 5.13

The trunk to limb ventral muscles in the upper limb (the pectoral muscle group) in non-human primates (a) and in humans (b).

size and complexity of this dorsal muscle group in upper limbs as compared with lower limbs. Further, in humans, there can sometimes be found any one of the many extra components of this muscle group routinely evident in non-human primates. In the human upper limb, however, comparison of the situation with that in the lower limb is more complicated. In the upper limb there are extra dorsal muscles from the head that are not represented in the hindlimb.

Thus one group at the shoulder not represented in the hindlimb comprises those individual head muscles of branchiomeric origin that have been drawn into the shoulder over both developmental and evolutionary time (most of trapezius and sterno-cleido mastoideus). These are very ancient evolutionary developments found in all land vertebrates (and indeed even earlier). It seems that, developmentally, these muscles and their nerve supply relate to a very ancient contribution of head somites and cranial nerves to the upper limb that is maintained in mammals and, therefore, humans. There is, of course, no representation of these head domain muscles in the hindlimb at all.

A second group of extra muscles at the shoulder are the rhomboid and serratus muscle sheets drawn from components that are embryologically and evolutionarily trunk derivatives. There is an equivalent muscle group in the hindlimb, the subvertebralis muscle of reptiles (the psoas minor muscle in mammals) but this hindlimb derivative is scarcely comparable with those in the forelimb where, in most mammals, there is no movable arrangement between the abdomen and the pelvic girdle as there is between the thorax and the scapula.

Also at the shoulder are limb muscles proper (the latissimus dorsi, teres major, dorsoepitrochlearis, and various parts of the deltoideus and triceps brachii muscles).

These muscles are also generally not represented in the hindlimb where the pelvis and sacrum are fixed, though the quadriceps femoris is a serial homolog of triceps (both these muscles cross the more distal joint, elbow and knee, respectively) in the two limbs.

Let us look at these various muscle groups in turn.

5.3.10 Head Derivatives in the Limbs: the Branchiomeric Muscle Sheet

Trapezius, a branchiomeric component, (we are excluding sternocleidomastoideus here because it does not cross the shoulder joint) in non-human species is usually heavy and has an extensive linear origin that can start as far laterally as the mastoid process of the skull, include the entire superior nuchal line or crest, the external occipital protuberance, the ligamentum nuchae (and through it, therefore, the upper 6 cervical spines), the 7th cervical spine and the spines of all the thoracic vertebrae, and sometimes as far caudally as the third lumbar vertebrae (Fig. 5.14(a); note: this figure also shows latissimus dorsi and its various components, see later).

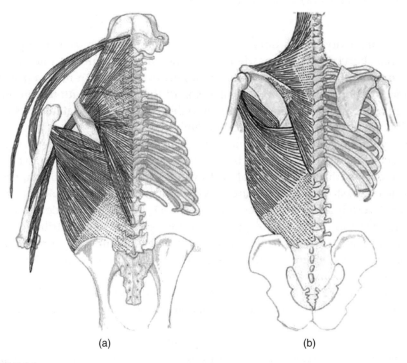

(a)   (b)

FIGURE 5.14

The branchiomeric muscle sheet in the upper limb (the trapezius muscle group) in non-human primates (a) and in humans (b). (This figure also shows the latissimus dorsi muscle group – see later.)

In humans, in contrast, the muscle is very thin; its upper limit is restricted to the medial third of the superior nuchal line; its lower limit to the 12th thoracic vertebral spine (Fig. 5.14(b)).

The insertion of the muscle likewise varies. In non-human species the insertion is extensive (Fig. 5.14(a)). The most superficial and most cranial fibers may pass into fascia of the arm and even forearm; other cranial fibers may insert into the region of the deltoid tuberosity on the humerus. On the clavicle the insertion can be as extensive as to include the lateral two thirds of the clavicle and again the fibers may be conjoined with the fibers of the anterior deltoid. Indeed, its most superficial fibers may be coextensive with the superficial layers of the anterior deltoid, thus forming a cephalo-humeralis muscle (regularly found in horses). This is present in many animals that have no clavicle, but it can also be found in some primates (all primates have clavicles), for instance in some lorisiformes).

In humans, again in contrast, the much smaller trapezius muscle has its insertion restricted to the lateral third of the clavicle, the acromion and the upper edge of the scapular spine and a small tubercle near the vertebral border (Fig. 5.14(b)). Variations of this muscle in humans, however, though usually quite small, nevertheless indicate that humans have the same propensity to develop the many extra components.

5.3.11 Trunk Derivatives to the Limbs: Rhomboid and Serratus Muscle Sheets

Two more sheets of dorsal muscle, possibly trunk derivatives that have migrated into the limb, comprise muscles that move the limb girdle upon the trunk, the rhomboids and the serratus muscle blocks.

Rhomboid Muscle Sheet

In non-human forms the rhomboids are extensive and have cranial (occipital) as well as complete cervical and thoracic heads. They originate from far laterally on the superior nuchal line extending medially to the external occipital protuberance, from the nuchal ligament at the levels of all cervical spines, from the spines of the last cervical vertebra and thoracic vertebrae often extending as low as the last thoracic vertebra in some species. Though basically a single sheet, their separately named origins mean that in non-human primates they are named rhomboideus occipitalis, cervicis and thoracis, respectively. They often show intermediate separations so that there can be as many as 6 or 7 apparently distinct muscles (Fig. 5.15(a)).

In humans, in contrast, the rhomboids are small and usually limited to two muscles, the rhomboideus minor and major of medical anatomy, originating from the lowermost end of the ligamentum nuchae (the seventh cervical vertebra) and the remaining spines down to the fifth thoracic (Fig. 5.15(b)). Yet, on occasion, in humans small cervical and even occipital heads occur as variations. When they occur they are surrounded by the same sets of fascial sheets as in non-human species. The fascial sheets are present even when the variations are not.

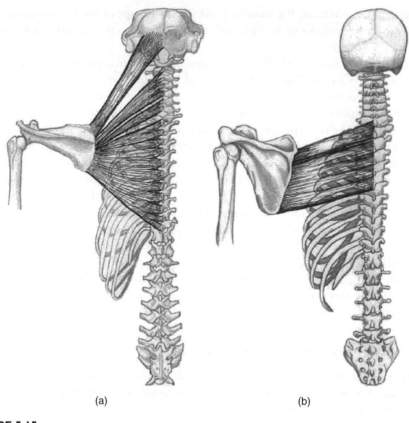

(a)   (b)

FIGURE 5.15

The trunk muscle sheets to the upper limb (the rhomboid muscle group) in non-human primates (a) and in humans (b).

The Serratus Muscle Sheet

The serratus muscle block also moves the limb girdle upon the trunk. In non-human forms the various serratus muscles, have extensive origins and insertions. They originate from lateral bony processes as far cranially as the mastoid process (sometimes named masto-scapularis or cranio-scapularis), from the lateral aspects of the superior nuchal crest (occipitoscapularis, often even two or more such muscles), from the transverse processes of the atlas (again, often, more than one muscle giving atlantoscapulares anterior et posterior), from all of the transverse processes of the other cervical vertebrae (a levator scapulae but much more extensive than in humans) and from the lateral aspect of ribs (serratus anterior vel magnus) often to as low as the eighth rib (and even as low as the thirteenth in some creatures such as bats). These muscles are separately distinguishable, hence the profusion of names. It is clear however that these are different degrees of development of a single muscular sheet (Fig. 5.16(a)).

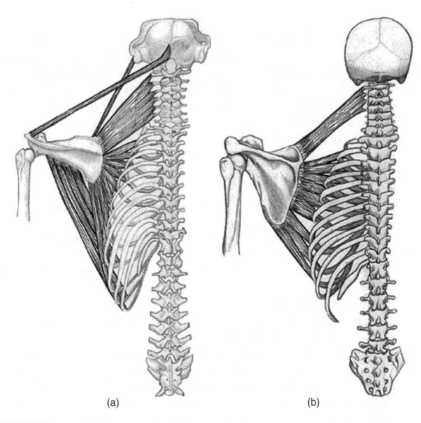

(a)   (b)

FIGURE 5.16

A second set of trunk muscle sheets to the upper limb (the serratus muscle group) in non-human primates (a) and in humans (b).

The insertion of this complex serratus group in non-human primates, and even more so in many other mammals, is likewise very extensive and includes as much as the lateral two thirds of the clavicle, the acromion, parts of the superior border of the scapula and the whole of the medial (vertebral) border of the scapula.

(In bats, the forms in which this muscle is most extensive, this insertion further includes almost the whole of the axillary border of the scapula. The additional arrangements in bats are presumably a further extension of the shoulder muscula- ture attendant upon the fact that in bats the scapula is not just a flat bone laid upon the dorsum of the thorax but, in fact, the equivalent of a long bone extending so far laterally away from the trunk that it forms a true first functional segment of the wing).

In humans, in contrast, this muscle sheet comprises only the two obvious muscles of standard human anatomy, a levator scapulae with an origin usually restricted to the first four cervical transverse processes and a serratus anterior with an origin restricted to the first to eighth ribs (Fig. 5.16(b)). Likewise in humans, the insertions of these two muscles are restricted to the medial (vertebral) border of the scapula.

Again, the fact that this has not always been the case in the human lineage is evident from the large number of recorded human variations. These include mastoido-scapular and occipito-scapular muscles, atlantoscapulares anterior and posterior, and levator claviculae muscles together with various muscle bundles linked with other associated muscles. All of these have their investing fascial sheets. Whenever these variations are present they are usually very small and would appear to have little functional importance. They may again, however, be useful indicators, ghosts, of past situations.

5.3.12 True Dorsal Limb Muscles

Finally, there are the dorsal muscles that act at the scapulo-humeral joint. These do have equivalents in the hindlimb.

In the hindlimb they comprise two blocks, one acting on the hip joint alone, and the other acting at both hip and knee.

The single joint group comprises the various gluteal and related muscles arising from the dorsum of the ilium, and the iliacus muscle arising from the ventral surface of the ilium (both sides of the ileum are dorsal components of the pelvis). In the forelimb the equivalent shoulder joint muscles are the latissimus dorsi and teres major, and deltoideus and subscapularis (not figured).

The two joint muscle block of the hindlimb comprises rectus femoris and sartorius, and the main one joint muscle components, the remaining parts of the quadriceps femoris. These are all embryologically dorsal muscles even though lying on the anatomically ventral surface of the thigh following limb rotation during development (as previously explained). The equivalent dorsal muscles in the forelimb are the various heads of triceps and dorsoepitrochlearis.

In the hip and thigh, these groups are large and complex in both non-human and human forms. It is true that part of the complexity relates in non-humans to elements of these muscles that move the tail (e.g. caudo-femoralis) and in humans to an increase in size of that portion of the gluteus maximus that inserts into the iliotibial fascial aponeurosis, band or tract. These seem to relate to functional differences between quadrupedal and bipedal movement. The quadriceps femoris in both non-humans and humans is large and complex, being primarily associated with knee extension, a critically important movement both in animals that use all four limbs in locomotion, and in humans that use only the hindlimbs in weight-bearing aspects of locomotion. In general terms, however, the relative sizes and degree of complexity of all these muscles in the lower limbs are similar in both humans and non-humans.

In contrast, in the shoulder and arm, these muscle units, for example, latissimus dorsi, deltoid and triceps, are much smaller in humans, though their complexity is not so very different. There are, however, in humans, along with these generally much reduced muscles, numerous small muscular and fascial variations, (including the aforementioned dorsoepitrochlearis muscle) reflecting those usually present in non-human primates.

5.3.13 Equivalences in Hands and Feet

It would be good to examine in a similar way, the equivalences in muscle structures in the distal parts of the limbs, the hands and feet. And in truth there is much information about them; they have been much dissected. However, it is possible to provide information that implies that similar differences are apparent between hands and feet, without going through all the many details of hand and foot muscles.

Thus the relative sizes of hands and feet give the information. If we compare, size by size, the dimensions of hand and feet in the various apes, we find that, given that they have somewhat different detailed structures, they are approximately equal (Fig. 5.17(a–c)). When, however, we come to make the same comparison in humans, we recognize immediately that there has been a great reduction in the human hand

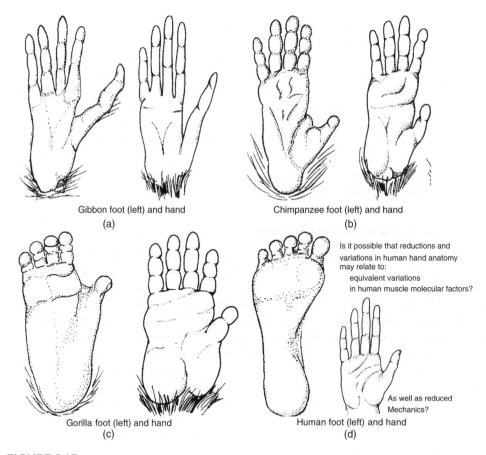

Gibbon foot (left) and hand
(a)

Chimpanzee foot (left) and hand
(b)

Gorilla foot (left) and hand
(c)

Human foot (left) and hand
(d)

Is it possible that reductions and variations in human hand anatomy may relate to:
  equivalent variations
  in human muscle molecular factors?

As well as reduced
Mechanics?

FIGURE 5.17

Feet, left, and hands, right of (a) orangutans, (b) chimpanzees, (c) gorillas, and (d) humans, all in their correct size relationships within each species.

compared with the human foot (5.17 (d)). This must, of course, be mirrored in the hand and foot musculature.

5.3.14 Conclusions for Muscles

It is evident from a great deal of information, therefore, that, despite small differences in the hindlimb musculature between non-humans and humans, there is a general concordance between them in size and complexity. The developmental processes that produce large complex muscular sheets in non-humans are also present in humans.

In the forelimb of non-humans the muscles are likewise also large and complex, and to this degree resemble the equivalent hindlimb muscles of both non-humans and humans. In unique contrast the forelimb muscles of humans are enormously reduced, both in complexity and size from what is found in the forelimbs of non-humans.

Yet humans have occasional variations that are similar, but much smaller, to many of the extra muscular elements that exist in non-human primates.

It is likely that the same molecular factors producing muscle components are present in both non-humans and humans. Is it possible that, in humans alone, some of their activity in the upper limb has been repressed or even extinguished in some way? Only in a few individuals, would it appear that non-human-like variations occur. Though this idea can only be tested by the work of molecular biologists, the likelihood of it being correct is supported by the findings in the skull. Here, where molecular factors are known to differ in humans uniquely in relation to jaw muscles, anatomical examination of jaw muscle variations provides the morphological evidence that goes with them (next).

5.3.15 Functional Implications for Bones

Reductions of the temporal muscles (see later) are associated with differences in the bones to which they are attached. These differences, reduced mechanical robusticity and changed bony surface features, are discernible in humans as compared with apes and monkeys. They might possibly be discernible in the best preserved of the fossils though this has not, so far, been examined. The reduction in the human temporal muscle is associated with evolutionary dates obtained from molecular studies. Thus, molecular phylogenetics has already given rise to assessments of the times at which reductions in the muscles of mastication in humans may have occurred (possibly between 2 and 3 million years ago for the MYH16 factor in the head domain). These dates are consistent with changes in the skull following experimental reduction of biomechanics of these muscles.

As in the skull, such reductions in human upper limb (but not lower limb) muscles could be associated with reductions in upper limb (but not lower limb) bones due to changed biomechanical loads.

For example, one important mechanical parameter of any long bone is the ratio of the diameter of the bone divided by the thickness of the cortical wall. This ratio (or equivalent) is a component of the second moment of area, a mechanical quantity that relates to bending. In creatures in which bending predominates over compression this ratio tends to be high (e.g. as high as 20 – 40) for example, in the forelimb bones of gliding dinosaurs where almost pure bending is produced in wings largely held stiffly extended in gliding Fig. 5.18). There is, presumably, a small component of bending due to off-axial loading of limb muscles when such a wing is 'flapped' but in contrast to wings in flying rather than gliding animals, 'flapping' is not a major component of such gliding flight.

In birds and bats that have true flapping flight and wing muscles that are large and powerful, the ratios are less, of the order of 10–14. In terrestrial quadrupedal animals the ratios are smaller still, varying from about 7–8 in small springy creatures like the smaller antelopes to as low as 3 in heavy, though still capable of fast movement, animals such as elephants and rhinos (Fig. 5.18). In these various creatures, in comparison with the flying and gliding species, the degree of pure bending is reduced as compared with increases in the component of bending due to off-axial loading of muscles, and increases in axial compression due to body weight (lower in antelopes, higher in elephants).

The ratio is actually at its minimum (2: i.e. no marrow cavity at all, hence the bone diameter is twice the wall thickness, which is the radius) in very slow, and presumably not at all springy, but very heavy creatures like the giant ground sloths of North and South America, and some giant Australian marsupials such as *Zygomaturus* and *Palorchestes* (Fig. 5.18). The assumption here is that, due to great weight and slow

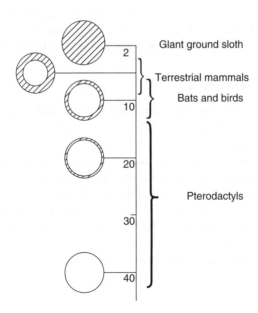

FIGURE 5.18

The ratio of bone diameter to wall thickness in the limbs of a series of creatures.

movement, bending is at a minimum in comparison with a maximum of compression produced by great body weight acting axially along a bony column in large very slow animals. (It is true, however, that we do not actually know for certain how these fossil and sub-fossil species moved.)

Among primates, the larger apes and monkeys have intermediate values with a ratio of bone diameter to cortex thickness (of around 4.0 to 5.0 depending upon animal size) for both forelimbs and hindlimbs. In these creatures, irrespective of differences in locomotor patterns, both limbs share fairly equally in the loads of locomotion though there is a slight tendency to a lower figure in hindlimbs.

Humans, in contrast, have quite different values for the two limbs: around 4.0 for the lower limbs but as high as 5.5 for the upper limbs. This ratio difference is likely due to the loss of general weight bearing by the upper limbs in creatures that move on two legs and do not use the upper limbs at all for weight bearing. This difference in upper-limb and lower-limb ratios of bones in humans may thus be a measure of the differential reduction in size and complexity of the upper-limb muscles in humans consequent upon bipedality.

This is also implied by old experiments in which muscular activity in limbs has been ontogenetically reduced. The effect is to greatly affect bony widths, thinning the wall of cortical bone, thus increasing the second moment of area, without much affecting bone dimensions.

These various bony architectural features are fairly readily obtained from fossil material. Thus general limb robusticity and the wall thickness/bone diameter ratios (where they can be measured or estimated) imply that australopithecines used both limbs in locomotion as do today's non-human primates, but that most early *Homo* species used only hindlimbs for locomotion. The reduction in upper limb muscles occurred separately in humans, and was not a feature of australopithecines.

5.3.16 Implications for Evolution

The standard view is that the many individual bone/joint/muscle arrangements in the phenotype that are mechanically efficient are adaptations that are acted upon by natural selection. To the degree that these are hereditary in nature, selection allows them to spread within a species. This may not, however, be the only way in which certain major episodes of human evolution have occurred.

Is it possible that the reduction in the forelimb domain muscles outlined here may be associated with some molecular alteration or alterations (almost certainly this may involve a cascade of factors rather than just a single factor) in the forelimb muscle system that, among primates, is confined to humans? This is a question for developmental and molecular biologists. Certainly a molecular alteration in relation to the muscles of mastication in the head has resulted in reductions in this muscle group.

Thus, in both the head and upper limb situations such molecular reductions would be enormously and immediately maladaptive in any individual non-human primate. As a result, they would be lost immediately they appeared. In other words, this is not a question of selection in populations and over time. For example, a reduction of masticatory muscles (in creatures primarily feeding on strong vegetable materials) or a reduction in forelimb muscles (in creatures habitually using the forelimb for locomotion) would be lethal at an early stage in every individual in which the change had happened to occur.

But in a creature in which aspects of mastication had changed (for example, less fibrous vegetable foods, softer animal foods, perhaps even the effects of food cutting, chopping, mashing, marination, even heating and cooking), or in a creature that had already become largely bipedal (for example not using its forelimbs for weight-bearing aspects of locomotion), such muscular reductions and losses in jaws and upper limbs would not be maladaptive. They might well, therefore, extend rapidly throughout such populations.

The occasional persistence of small remnants of many of the original muscular heads in upper limbs (and similarly in the masticatory muscles of the head) that have been generally lost in humans may be paradoxical indicators of these losses. Likewise, the continued presence of fascial sheets that are associated with the more extensive muscular arrangements in non-humans, may, even when the muscles have disappeared, be, likewise, indicators of the prior condition. These fascial retentions, in both the upper limb and the head, would stem from the fact that connective tissues are derived from connective tissue factors, and not the myotomal factors that influence muscles. In other words, the fascial sheets remain because they are not affected by the changes in the molecular factors that result in reduced muscles.

Is it thus possible that a further series of molecular studies could provide information about the evolution of bipedality? The definitive answers to these matters will surely lie with further findings in developmental biology. Almost certainly they will be more complex than just a single molecular event. This is, however, an example of how understanding anatomical forms and patterns in evolutionary biology can throw up hypotheses for testing in developmental biology, and of how findings from developmental molecular phylogenetics may be tested by the structures of anatomy. Wouldn't it be fascinating if the evolution of the human type of bipedality were as evident from the losses sustained by forelimb regions, as from additions in hindlimb regions (such as the hip and foot) - the usual evidences for bipedality?

CHAPTER SIX

Understanding the Human Head

6.1 Insights from Building the Trunk

As we have seen, the formation of the comma-shaped embryo naturally leads on to the more human-like fetus. The trunk in the fetus consists of two fundamental components: the primarily segmented external trunk interacting with the organism's external world – sensory and motor functions – and the largely unsegmented internal trunk concerned with the functioning of the organism's internal world – vegetative and visceral functions. (The limbs, also involving external functions, are merely a special ventral formation at certain points along the external trunk).

The Scientific Bases of Human Anatomy, First Edition. Charles Oxnard.
© 2015 John Wiley & Sons, Inc. Published 2015 by John Wiley & Sons, Inc.

Following this pattern, the development of the head, and therefore its final adult arrangement, should, in theory, also be describable through the same external and internal elements. Indeed the head does show such similarities to the trunk.

Thus, in parallel to the somatic external trunk, there is a somatic external head that has to deal with the external world. As a result, therefore, there are some structures in the head that are organized like the segmented somatic structures we have already recognized in the trunk.

However, perhaps because the head goes first into the environment, and perhaps also because the head may be in evolutionary terms an earlier structure than the trunk, there exists a whole series of additional special sensory structures to handle these first and most important external contacts. These extra structures provide us with information about chemicals, radiations, sounds, positions, and movements stemming from the environment.

Likewise, in parallel to the visceral internal trunk, there is a visceral internal head that has to deal with visceral functions in relation to the internal world of the organism.

Again, however, because the head goes first into the environment, there are additional special structures in this visceral internal head. These have to handle what might be thought of as visceral functions resulting, again, from that first entry into the organism from the environment. Thus they deal with the entry of various materials important for internal functions, for example, solids, liquids and gases, from the external world. As a result, though the visceral functions associated with the internal trunk are carried out by largely unsegmented trunk structures (the gut and its related parts), the internal head has come to have additional special visceral functions that are carried out by specially segmented visceral structures (the branchial or pharyngeal arches, clefts and pouches) leading to the gut tube in the trunk. They handle the intake and preparation of food and fluids for the more caudal gut tube, the intake and output of the respiratory gases for a caudal gut tube derivative, the respiratory system, special sensation and movement related to the above, and much else besides.

These differences between the structures of the trunk and head are associated with equivalent differences in the peripheral nerves subserving these developments.

There are, therefore, in the head, nerves (somatic sensory and somatic motor components) that are like the spinal nerve roots in the trunk. However, in the head they remain separate, as separate cranial nerves.

The additional special sensory structures that specifically examine the first entry into the external environment are supplied by additional special sensory nerves (also called cranial nerves, even though some of these are really tracts of the brain reaching out to the true peripheral nerve cells located within the sense organs). Likewise, the additional segmented special visceral structures (branchial (pharyngeal) arches and related structures) are supplied by additional special visceral (branchial, pharyngeal) nerves. These differences are summarized in Fig. 6.1.

Paradoxically, though the above discussion indicates that the head is more complex than the trunk, elements of these complicated head patterns (head somitomeres)

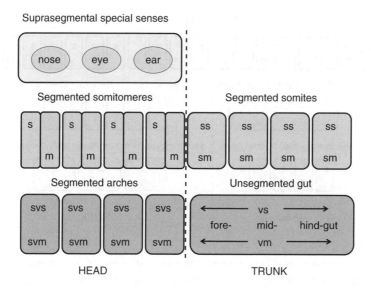

Suprasegmental special senses

FIGURE 6.1

Diagrammatic representation of segmentations in the different elements of trunk and head. In the trunk, SS and SM indicate somatic sensory and somatic motor components of each somite, and VS and VM indicate visceral afferent (sensory) and viscero-motor (autonomic motor) nerves of the viscera. In the head, S and M indicate somatic sensory and motor components of alternate somitomeres, and SVS and SVM special visceral sensory and motor components of branchial arches.

hark back to very early developmental complexities (trunk somitomeres) that, in the human trunk, disappear even before their formation in the human head. These same elements also hark back to structures in very early evolutionary ancestors, reflected today in separate segmental structures for sensory and motor components in much simpler species such as amphioxus. In other words, both developmentally and evolutionarily, earlier trunk structures that were later eliminated have, in the more complex head, been retained, for functions related to the head going first into the environment Fig. 6.2.

Another difference relates to the gut tube. In the trunk it is not segmented, but in the head the comparable component, the oro-pharynx, is segmented, giving rise to pharyngeal (or branchial) segmentation. And this, well recognized in the head, is not present in the trunk (Fig. 6.3) except in a few of the very simplest creatures (e.g. amphioxus).

Thus in the trunk we can relatively easily separate, as a concept, the external trunk, with its largely external functions of sensation and movement, and its largely somatic origins, from the internal trunk, with its largely unconscious and involuntary functions, and largely visceral origins. In the head, in contrast, though these two components do exist, there is much greater and more complex interaction between

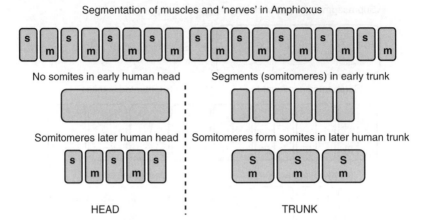

FIGURE 6.2

Diagrammatic representation of segmentation differences in somatic structures of head and trunk in the change from early to later human embryos, and in comparison with adult Amphioxus. The symbols are the same as in Fig. 6.1.

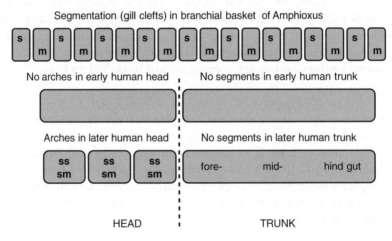

FIGURE 6.3

Diagrammatic representation of segmentation differences in visceral structures of head and trunk in the change from early to later human embryos, and in comparison with adult Amphioxus. The symbols are the same as in Fig. 6.1.

them, together with yet other complexities. As a result, though for many years it was the custom to separate trunk from head as we have done in the chapter headings, in reality this is more complex. The separation between head and trunk can now be seen to be at different levels.

A first separation is that achieved during beheading or the guillotine; but of course that is truly simplistic. That is, the head is separable from the trunk by the division at the neck between the body and the head, between the vertebral column and the skull, and between the spinal cord and the brain. However, there are also other levels of separation between the head and trunk.

Thus a second level notes that the caudal head muscles (e.g. the tongue), the caudal head bones (e.g. the occipital), and the caudal part of the brain (the hindbrain) have, like the trunk, somites (occipital somites) whereas everything rostral to them has somitomeres. This implies that the true separation is not at the caudal limit of the skull and hindbrain, but within the skull at the junction of the hindbrain with the midbrain. (This is also implied by the fact that the antecedents of the upper limb, normally thought of as a trunk derivative, have elements that stem from occipital somites and cranial nerves of the hindbrain).

A third level of separation relates to the division between the hind- and midbrains taken together because they are defined by the notochord, and the forebrain that is rostral even to the notochord. This is implied because one can think of the notochord being what defines trunk.

All three of these concepts are real and have become more obvious as more has become known about the developing central nervous system and the degree to which it influences, and is influenced by, the more peripheral structures.

We will, nevertheless, continue to separate trunk from head, but throughout we will attempt to show the importance of these various levels of division, both between the trunk and the head, and within the head itself.

6.2 Now into the Head

6.2.1 Building Components

The head, thus, contains some of the same building components as the trunk. These are, from superficial to deep: the superficial **cranial ectoderm**, the middle **cranial mesoderm** in two portions, **paraxial** and **lateral plate**, and the deepest **cranial endoderm** and **notochord** (Fig. 6.4(a)).

In the head, however, the conversion of some ectoderm into a **neurectodermal** structure (as in the formation of the spinal cord in the trunk) is more complex. This is due to the development of complex conjoint structures from two or even three of the above building components (Fig. 6.4(b)).

One comprises a series of bilateral dorsally placed **ectodermal/neurectodermal placodes** relating to the cranial special senses (Fig. 6.4(b)). These are places where external ectoderm and internal neuro-ectoderm (from the neural tube) lie in contact in the building of special sense organs (such as the olfactory and optic organs).

A second consists of a series of more ventrally placed **pharyngeal (branchial) placodes** relating to developments of special structures concerning the head functions of the beginnings of digestion and respiration (Fig. 6.4(b)).

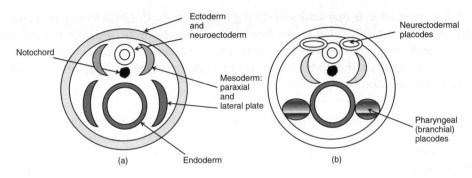

FIGURE 6.4

(a) Structures in a diagrammatic cross-section of developing human trunk showing major components. (b) The essentially similar pattern in the early head development with the addition of neurectodermal and branchial placodes.

A third is a cranial midline **ectodermal–endodermal complex** relating to the junction of the ectoderm and endoderm without intervening mesoderm at the cranial end of the embryonic head. This relates to the eventual cranial openings of the gut tube (mouth, nose).

A final one is the special region, **neural crest**, which commences as the conjunction of the cutaneous ectoderm with the neural ectoderm. This latter conjunction, which soon separates from both the progenitors of the skin and the brain, is similar in origin to the neural crest in the trunk and it later migrates, as it does in the trunk but much more extensively.

Let us take each of these in turn.

Head Modifications of Trunk Components

Ectoderm
One part of the ectoderm (**cranial ectoderm**) covers the entire external surface of the head and eventually becomes the outer layers (epidermis) of the skin of the head. It is contiguous, of course, with the ectodermally-derived skin of the trunk. The other part is a component (again as in the trunk) that invaginates in the midline to form a neural groove, and then a neural tube, along the midline of the embryo separated from the external skin (cross-section in Fig. 6.4(b)). However, in contrast to the situation in the trunk, in the head this component (**cranial neural tube**) becomes enormously enlarged, forming most of the brain. The brain is separately described in Chapter 7.

Mesoderm
One part of the mesoderm (**cranial paraxial mesoderm** or **somatic mesoderm**) lies on each side of the cranial end of the notochord. It shows some segmentation, and parallels, to some degree, the equivalent segmented paraxial mesoderm of the trunk.

A second part of the mesoderm, sometimes called **cranial lateral plate mesoderm,** or **visceral mesoderm,** is also called **pharyngeal** or **branchial mesoderm.** This lies more ventrally and laterally to the cranial end of the endodermal tube. It is somewhat similar to the plate mesoderm of the trunk. In the head, however, it is more complex and becomes segmented (pharyngeal or branchial arches). Further, it does not develop a serous cavity within itself as in the trunk (cross-section in Fig. 6.4(b)).

Endoderm
The **endoderm** lies ventral to the midline ectodermal neural tube and notochord, also as in the trunk. It forms the cranial portion of the gut tube and its derivatives. It is, again, more complex than in the trunk (cross-section in Fig. 6.4(b)), linking eventually with head ectoderm without any intervening mesoderm. It has elements of segmentation related to that of the developing pharyngeal (branchial) arch system.

Notochord
The **notochord** is the original skeletal axis of the embryo that is also present in the trunk. As in the trunk it lies between, on the one hand, the ventral gut tube (in the head the ventral pharyngeal precursors) and on the other, the dorsal nerve tube (the dorsal brain precursors, cross-section in Fig. 6.4(b)). As in the trunk, the notochord largely disappears during later development, but its developmental effects on the other structures can still be discerned. In the head, however, it does not extend all the way to the cranial end so that the cranial-most portion of the head has no notochordal element.

Neural Crest, a 'Fourth Germ Layer'
Though also present in the trunk, and associated in the trunk with the formation of a number of important structures (e.g. peripheral nerve ganglia and the sympathetic system), the **neural crest** becomes associated, in the head, with many other developments related to the greater complexity of the head, and especially to the aforementioned unique head segmentation, the branchial system. The neural crest is so much more complex than in the trunk that it has been described as a fourth germ layer. It is a critical elaboration in the head that is described later.

Summary
These several building components are, as indicated above, modified in the head in ways different to those in the trunk. Not all head developments, however, are more complex than in the trunk. For example, in the trunk a serous cavity develops within the visceral mesoderm. This does not occur in the head. Again, one portion of mesoderm found in the trunk, the intermediate mesoderm, is present only very transiently at the caudal end of the head. It gives rise to vestiges of the pronephros (already described with the internal trunk). These rapidly disappear so that the head later has no intermediate mesoderm.

With these *caveats*, we can continue with additional elaboration of these fundamental components that do not occur in the trunk.

Head-specific Differences in Building Components

Cranial Ectoderm

One part of the ectoderm, that covering the entire external surface of the head, is organized, as in the trunk, into overlapping dermatomes in relation to its general somatic afferent nerve supply. This nerve supply is derived, however, not from a series of spinal nerves, but from the general somatic afferent components of a nerve that is called the trigeminal, **Vth**, cranial, nerve. There are reasons for thinking that this nerve supply to the ectoderm actually comprises more than a single segment even though these somatic sensory fibers are all carried in a single nerve.

Another part of the ectoderm (neurectoderm) that (as in the trunk) invaginates in the midline to form a first neural groove, then a neural tube, becomes, in the head, enormously enlarged, forming the brain. This shows first three primary enlargements: the forebrain or prosencephalon, the midbrain or mesencephalon, and the hindbrain or rhombencephalon (Fig. 6.5). A third component, lying between the dermal ectoderm and the neurectoderm is, as in the trunk, the neural crest. This is, however, much more complex than in the trunk and is considered later.

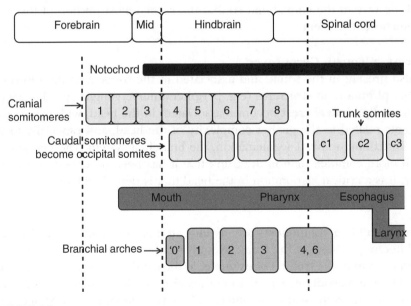

FIGURE 6.5

Elaboration of Fig. 6.2 showing more complete numbers of segments in head and more detail of transition to trunk.

Cranial Paraxial Mesoderm

The paraxial mesoderm in the head, in contrast to that in the trunk, can be considered in three parts.

The first is cranial-most (or rostral, the term rostral is often used in the head) and forms a mesodermal block that is called the **prechordal mesoderm** (because it is cranial to the cranial limit of the notochord). For this reason alone, it is already different from the situation in the trunk. It becomes segmented into somitomeres like those that existed transiently in the very early trunk. Unlike the situation in the trunk, however, these do not combine in pairs into somites.

The second part is caudal to this and lies alongside the cranial part of the notochord (or axis) and is, therefore, called **notochordal paraxial mesoderm.** This portion also segments into somitomeres and, again, these do not unite to become somites as in the trunk.

The third part, lying even more caudally is called **occipital paraxial mesoderm** and, in this part of the head, some of the initial pairs of somitomeres do unite, thus becoming more like the somites of the trunk. For this reason these structures are also called **occipital somites** (occipital because they are related to the caudal-most portion of the skull, the occipital bone).

These various separate designations of the mesoderm are not just for descriptive purposes. They relate to their different eventual fates, and emphasize both similarities and differences between the head and the trunk (Fig. 6.5). For example, one could think that it is at the level of the rostral end of the notochord that the real trunk begins. However, another division between head and trunk might be at the changeover from somitomeres to somites. Finally, it is yet also possible, from the derivation of the occipital somites, that their caudal end is where the real trunk begins. Consideration of the nervous system development and structure (Chapter 7) indicates that this is even more complicated; however, this is dealt with later. These various arrangements further explain why the separation into head and trunk is somewhat artificial.

Cranial Lateral Plate Mesoderm

In the head the lateral plate mesoderm resembles that in the trunk in lying largely ventral to the notochord and the developing nervous system, and, therefore, alongside (hence lateral) to the developing gut tube (here to become the pharynx) and its other eventual derivatives (larynx and upper part of the trachea). It differs, however, from the situation in the trunk in several ways.

First, the lateral plate mesoderm is not divided into superficial and deep portions with an intervening serous cavity as in the trunk. It remains a single solid component as the **cranial lateral plate mesoderm,** and is also known as the **pharyngeal (branchial) mesoderm.**

Second, unlike in the trunk, it becomes segmented. This segmentation gives rise to the material within what come to be called the **pharyngeal (branchial) system** of **arches** and between them the **pharyngeal (branchial) system** of **clefts** in the overlying ectoderm and **pouches** in the underlying endoderm. (In many aquatic vertebrates, the structures are actually open to the exterior (like the mouth and nose) as gill slits).

Though in a few animals (tunicate larvae, lampreys) there are a large number of such arches (gill slits) that lie not only in the head but along most of the trunk, there are many fewer in most vertebrates and they lie only in the head. In some there may be as many as nine. In humans there are only four (some are thought to have disappeared during evolution). There are also some materials in humans that may represent an original first arch (see below). Because the system of numbering was decided before these remnants were discovered, this arch, if it is recognized at all, might be termed the '**0th** arch' (Fig. 6.5).

The human first (**mandibular**) and second (**hyoid**) arches are well defined. The remaining more caudal arches, show elements of coalescence, and are often grouped as simply **pharyngeal (branchial) arches**.

Cranial Endoderm

The endoderm in the head, lying ventral to the ectodermal neural tube and the notochord, is the cranial equivalent of the endoderm in the trunk. Again, it differs from its arrangement in the trunk because it is partly segmented, in line with the aforementioned segmentation of the lateral plate mesoderm of the branchial arches. It comes to form the epithelial part of the cranial end of the gut tube, the pharynx and its alimentary and respiratory derivatives.

Cranial Notochord

The notochord largely disappears, as in the trunk. It is, however, as in the trunk, fundamental to the beginnings of the early building processes.

Further Head Complexity: New Double and Triple Components

There are further complexities of the building materials in the head that differ totally from the simpler situation in the trunk. These involve local interactions between two or three of the building materials. They include: ectodermal–endodermal complexes, ectodermal–ectodermal complexes, and ectodermal–mesodermal–endodermal complexes (See earlier and Figs. 6.4(a,b).

Midline Ectodermal–Endodermal Complexes

These are regions where the endoderm and ectoderm lie together without intervening mesoderm.

One of these is in the midline where the cranial end of the internal gut tube eventually opens out onto the external skin surface. This complex is thus composed of two interacting building materials, ectoderm and endoderm. They form the junctional epithelium between the external cranial skin and the internal gut mucous membranes. This junction is also called the **oral plate** and when this plate breaks down, so that skin epithelium becomes conjoined with gut epithelium, it forms the actual openings to the gut (mouth, nose). (Similar, but much smaller developments, occurs in the trunk, at the region of the exit of the various umbilical structures ventrally on the abdomen, and at the hind end of the gut tube where the endodermal and ectodermal sheets are again in contact without intervening mesoderm thus forming a **cloacal plate** (which likewise, later, breaks down to form anal, urinary and genital openings) and see later).

Ectodermal–Neurectodermal Complexes

Another set of complexes are three pairs between **external ectoderm** (skin) and **internal ectoderm** (neurectoderm) without intervening mesoderm. They, too, do not exist in the trunk. Mentioned earlier, they are called **special sensory placodes**. These placodes do not break down (like the oral plate) so continuities between the inside and outside of the embryo do not develop at these structures. They are, nevertheless, regions where the external world comes very close (sometimes less than 1 mm away, e.g. at the olfactory mucosa) from the internal brain (with consequent important clinical implications). As their name indicates, these junctions are between portions of the developing brain and the overlying developing skin. They foreshadow the developments of different structures of the special sense organs: first, olfactory – smell, second, optic – sight, and third, stato-acoustic – balance/hearing (together with the zeroth special sensory modality described later). Though first dorso-lateral to the developing brain, they rapidly become covered over by the enormous dorsal growth of the brain.

Special Triple Complexes, Ectoderm–Mesoderm–Endoderm

Yet other regions of the head show triple relationships. These are complex structural developments including all three layers. These are the **pharyngeal (branchial) structures**, and are described in more detail below.

These, then, are the building components for the formation of the head.

6.2.2 Building Processes

These components undergo a variety of building processes that include: **longitudinal segmentations**, **transverse dorsoventral divisions** and various **longitudinal** and **transverse migrations**. Because of the additional complexities of the building components, these building processes are more complicated than in the trunk.

It is easiest to consider the building processes first in terms of each of the primary building components, ectoderm, mesoderm, endoderm, and notochord. Then we can consider interactions in the more complexly organized building components.

Building Processes in the Four Primary Components

Ectoderm

The ectoderm appears to have segmented dermatomes that seem not unlike those associated with each developing spinal nerve in the trunk. In the head, however, there are only three major 'dermatomes', the ophthalmic, maxillary, and mandibular regions of the face. Though sensation in these major 'dermatomes' is called somato-sensory (*vide* somato-sensory in the trunk) these areas are really also partly dermal areas of branchial arches. They are supplied by the ophthalmic, maxillary and mandibular branches of the trigeminal nerve. Though this is a nerve of a branchial arch, some of the fibers within it that supply the dermis are the equivalent of somato-sensory fibers in the dorsal root of a spinal nerve. Unlike the situation in the trunk, these fibers never join with their motor equivalents to form a spinal nerve equivalent). Further, the ophthalmic branch of the trigeminal may in fact be a remnant of the somatic sensory nerve and skin area of the elusive zeroth arch.

There are also equivalent small skin areas within and around the ear that are supplied by small equivalent general somato-sensory branches of each of the other branchial nerves, the facial, glossopharyngeal, and vagus.

In addition, a large part of the cranial skin, that is, almost all of the parts dorsal to the skull vertex in the midline and the ears laterally, comprise the first dermatomes from the trunk, and are supplied by dorsal branches of the first spinal nerves.

Mesoderm

The cranial **prechordal mesoderm** is segmented into somitomeres numbered 1–3 (Fig. 6.5). These somitomeres are located cranial to some of the developments described below. Though not forming somites, they still produce some general somatic structures, for example, the external **muscles of the eyeball** itself. They are supplied by cranial nerves (3, 4 and 6) that are the equivalent of the ventral roots in the spinal cord.

The middle **parachordal mesoderm** is segmented into somitomeres numbered 4 to 8 (Fig. 6.5). These produce structures in the segmented branchial arches. Thus they give rise to some of the muscles of the branchial (pharyngeal) arches, some of which eventually come to lie in the neck. The products of this segmentation are greatly enlarged by the addition of populations of cells (neural crest cells) that stream ventrally from the very dorsally located neural crest. These cell populations give rise to many of the cartilaginous elements of the branchial (pharyngeal) arches. These structures are not somatic but branchial even though they contain striated muscle (unlike the equivalent mesoderm in the trunk). The nerves that supply these muscles are thus not like trunk ventral roots, but are something special (hence special motor fibers) in the particular cranial nerves that supply the branchial arches.

The hindmost **occipital mesoderm** is initially segmented into a set of somitomeres and these do unite to form somites (as in the trunk). They are often called **occipital somites** (Fig. 6.5). They lead backwards into the cranial-most somites of the trunk, the cervical somites, with which they are quite similar. But unlike the cervical somites that remain separate at each segmental level, these occipital somites fuse cranio-caudally. The occipital somites give rise to myotomes that migrate very ventrally to form the intrinsic muscles of the tongue. They give rise to sclerotomes that form part of the base of the skull (the bodies of the sphenoid and occipital bones) and the bony surrounds of the foramen magnum (the hole in the skull base through which the spinal cord exits). Though in the adult these muscles and bones are apparently single structures, their development and nerve supply (the several roots of the 12th cranial nerve, again resembling ventral motor nerve roots of spinal nerves in the trunk) indicate their multiple segmental origination. In other words, segmentation in this region is partially hidden, and the precise number of segments is ill-defined.

Endoderm

The endoderm in the head involves, as we have seen, the epithelial linings of a tube with originally a cranial blind end, and leading caudally into the gut tube of the trunk. The cranial end abuts against the ectoderm with no intervening mesoderm, thus forming a double layered structure, the oral plate, as previously described.

Most of the tube caudal to the oral plate becomes segmented and in this it differs from the unsegmented gut tube in the trunk. The segments are evident as a series of lateral outpocketings from the endodermal tube on each side called **pharyngeal (branchial) pouches** (that become a number of important adult structures, see below). They are related to an equivalent series of clefts (inpocketing) from the overlying ectoderm called **branchial clefts** (also giving rise to adult structures, see below). The materials of the mesoderm are present between them in the branchial arches (Figs. 6.5 and 6.6). In aquatic forms the clefts and pouches are open, forming gill slits (hence the term branchial). This does not occur in mammals (therefore humans), hence the term pharyngeal.

There are also some midline, therefore single, outpocketings. One is an unpaired ventral midline structure (better, a diverticulum) called the **sub-pharyngeal groove.**

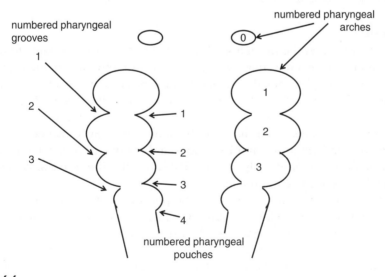

FIGURE 6.6

View from the dorsum of diagrammatic coronal section along the length of the branchial region in developing human showing main components.

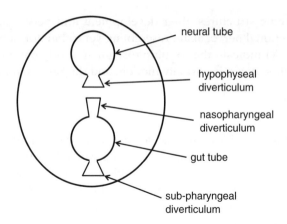

neural tube

hypophyseal
diverticulum

nasopharyngeal
diverticulum

gut tube

sub-pharyngeal
diverticulum

FIGURE 6.7

Diagrammatic cross-section of the development of midline structures from the neural and gut (branchial) tubes.

This becomes, among other things, the adult thyroid gland lying ventrally and migrating caudally so that it ends up in the neck. Another is an unpaired dorsal outpocketing, called the **naso-pharyngeal diverticulum.** This becomes the adult adenohypophysis that migrates dorsally and meets up with an equivalent ventral outpocketing from the base of the brain, the **neurohypophyseal diverticulum.** These two together become conjoined to form the doubly-organized adult **pituitary gland**. These particular developments are summarized in Fig. 6.7.

Notochord

We already know that the **notochord** is a structure that gives its name to the entire group of animals called the chordates that include not only all vertebrates, but also a number of even simpler animals, craniates and acraniates. It is a cellular rod running along the longitudinal axis of the embryo just ventral to the developing central nervous system. Although it has largely disappeared in the adult, it plays an extremely important role in a series of mechanisms that induce very early unspecialized embryonic cells to transform into definitive tissues and organs. In particular, in the head, as well as in the trunk previously described, it stimulates the conversion of overlying ectodermal tissue into neural tissue.

In the head the notochord is limited to the portion caudal to the developing forebrain and thus contrasts with the trunk in which it extends throughout its length. Within the developing brain and spinal cord it stimulates the conversion of some cells in the developing spinal cord (trunk) and brain (head) to become the most ventral and caudal parts of those structures (the **floor plates** – see brain and spinal cord, Chapter 7).

Building Processes in the More Complex Double Structures

Next we can consider the building processes in terms of the complex double and triple structures.

The Ectodermal–Endodermal Complex

This is called the **oral plate** and it changes by eventually opening up to form the cranial orifice of the gut tube. In so doing the orifice becomes divided by the ingrowth of two structures (a horizontal one, the eventual palate, and a vertical one, the eventual nasal septum, see below). These divide the original single opening into a ventral mouth and two dorsal nares, leading into the ventral buccal and dorsal nasal cavities. The ectodermal elements stream inwards so that many of the internal buccal and nasal tissues are also of ectodermal origin.

The Ectodermal–Neuro-ectodermal Complexes

These also undergo a series of changes. They arise as a result of inductions between neural ectoderm of the developing brain and the cutaneous ectoderm overlying it (together with some structural contributions from local mesoderm).

The most obvious are the paired sets of **placodes** involved with the three sets of special senses: smell, vision and balance/hearing that were briefly described earlier. These arise in relation to sensory extensions of the brain. There are also a series of other **placodes** in the branchial arches. These, too, are related with: sensing aspects of the external world such as taste, functions to do with the internal milieu, and the formation of related sensory ganglia of cranial nerves. The development of the pituitary gland involves a dorsal midline placode developing in concert with a ventral neural outpouching. These are each described separately below.

The 'Special Sensory' Placodes

These are associated with equivalent **sensory capsules**. Together they give rise to the sensory and supporting structures.

It is usually accepted that the most cranial (first pair) are **olfactory**. They are involved with the **sense of smell** and the brain extension called the **olfactory tract and bulb** (also previously known as the **Ist cranial nerve, the olfactory nerve**).

The second pair of placodes and capsules (**optic**) form more caudally giving rise to certain structures of the eye (e.g. the **lens**) and are associated with brain extensions to the eye called the **retina** and the **optic tract**. The optic tracts are also known as the **IInd cranial nerve**, the **optic nerve**. All of this is related to the **sense of vision**.

The third pair of placodes and their capsules (**otic**), arise more caudally still. They give rise to the many of the structures of the inner ear for the senses of **hearing and balance** and are supplied by nerve tracts back towards the brain, known as the **VIIIth cranial nerve**, or the **stato-acoustic nerve**.

Together these three developments are often called the special senses.

There is also a pair of very small special sensory placodes located even more cranially than the olfactory placodes. These are the **vomero-nasal placodes** that give rise to the vomero-nasal organs and their brain extensions, the **terminal nerves**. The sensory modality involved here seems like olfaction. It is, however, aimed at the chemical sensing of pheromones and was long unrecognized in humans. If olfaction is the general sensing of a multitude of chemical signals, mainly from the external world, and the business of the **Ist** cranial nerve (the traditional numbering system), then the terminal nerve is involved with special signalling related to individual chemical sensing, and the nerve has to be labeled the **0th** cranial nerve.

A Placode Relating to the Internal Milieu
This has already been noted (see Fig. 6.7). It is a very cranially placed midline **placode** that is small (compared with the main sensory placodes). This contributes to the developing **pituitary gland**. Its neural component migrates ventrally from the brain (hypothalamus) to become the **neurohypophysis** of the pituitary gland. It meets up with a second portion migrating dorsally from the epithelium in the nasopharyngeal region to form the **adenohypophysis**. In one sense these two portions of the pituitary have separate origins and merely undergo migrations that happen to result in their being in contiguity in the adult. In another sense, however, this anatomical contiguity is closely related to complex functional interactions between them. The eventual pituitary gland is thus a single architecture with two origins and a complex of important hormonal functions. The entire structure comes to lie within the skull, though a number of embryological deformities (open channels, small cysts, etc.) indicate the route through the skull base followed by the dorsal migration of the adenohypophysis during development.

Placodes Relating to Pharyngeal (Branchial) Segmentation
The first of the series of placodes arising laterally associated with pharyngeal (branchial) segmentation are, from cranial to caudal, the **paired trigeminal placodes** that become eventually the part of the sensory ganglia and associated structures for the main part of the **trigeminal nerve** (the so-called **Vth cranial nerve**). This nerve subserves most of the **sensation of the skin** of the head. It is, however, complex, being also associated with the first pharyngeal (branchial) arch.

Next in the series are a set of paired laterally placed placodes known overall as the **epibranchial placodes**. They are associated with sensory functions in the developing pharynx. Their functions are mostly special sensory to the various components of the pharyngeal branchial arches. Their serial arrangement is associated with the serial arrangement of these arches (of which the mandibular was merely the first). They thus form various elements of those arches: hyoid (second arch), and pharyngeal (third and fourth arches) and are supplied by specific components of the cranial nerves (**VIIth** (facial), **IXth** (glossopharyngeal), and **Xth** (vagus])) that supply these arches (Fig. 6.8).

The most cranial pair of these epibranchial placodes produces the neurons and ganglia of the **facial nerve**, the so-called **VIIth cranial nerve**, the nerve of the

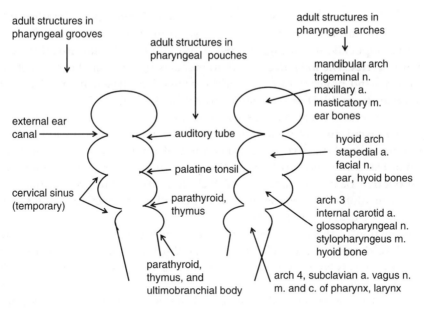

FIGURE 6.8

Same diagrammatic view as Fig. 6.6 coronal section along the length of the branchial region in developing human showing detailed adult derivatives.

second, hyoid, arch. This is involved in the **sense of taste**. The next pair of these produces the neurons and ganglia of the so-called **IXth cranial nerve**, the nerve of more **caudal arches**. (However, the numbering of the arches more caudally is less well understood in mammals.) These are associated with more caudal elements of the **sense of taste**. The third pair contributes to ganglia and neurons related to the **Xth cranial nerve**, also involved with more **caudal arches**. Both of these nerves, in a complexity that is very ancient in vertebrate evolution, include neurons that pass far caudally into the trunk supplying the **heart** and **lungs** in the thorax, and the **stomach and duodenum**, and many other parts of the more cranial viscera of the abdomen.

Each of these cranial nerves has proximal and distal ganglia. It is the distally placed ganglia that are derived from the placodes. The more central ganglia for each nerve are derived from a different source, the migrations of that ubiquitous neural crest (and are more like the dorsal nerve ganglia described for the trunk). Yet again, therefore, interactions are complex, but it can be seen that the whole arrangement has a logic that relates to development and evolution.

Building Processes in the Even More Complex Triple Structures

The Ectodermal–Mesodermal–Endodermal Complexes

These, as their name indicates, contain all three components. **Endodermal** parts bud outwards from the pharynx as a longitudinal series of internal outpouchings.

These are the paired lateral **pharyngeal (branchial) pouches. Ectodermal** components derive, on the surface of the embryo, from a series of segmented external ectodermal inpocketings (invaginations). These are called **pharyngeal (branchial) grooves** or **clefts**. The **mesodermal** components lying between these sets of pharyngeal grooves and pouches form a series of segmented mesenchymal masses called **pharyngeal (branchial) arches** (summarized in Fig. 6.8).

Some of these mesenchymal masses are from the somitomeres 4–8 and form the **incipient musculature** of each arch. Most of this mesenchyme includes, however, the aforementioned massive ventral migration of populations of neural crest cells from their extreme dorsal position on the neural tube. These cell populations form the **connective tissue** components of each arch. These are eventually associated with the formation of the **cartilages**, **bones** and other **connective tissue structures** (e.g. blood vessels etc.) of each arch.

The formation of these arches, numbered from the first (**mandibular arch**), the second (**hyoid arch**), to, less clearly obvious, third and fourth arches (**pharyngeal arches**) are closely associated with molecular factors involved in the developing brain that we will examine in Chapter 7.

It can thus be seen that this segmented system of arches is developmentally complex and gives rise to a wide variety of structures in the head and neck (See Figs. 6.8–6.10 and 6.11). Understanding this complexity is central to understanding (rather than simply memorizing) the structure of the adult human head. In addition to the developmental changes, some of this complexity relates to both very ancient evolutionary origins (even before the evolution of craniates as a group) and to more recent evolutionary modifications (in evolution from pre-reptiles to pre-mammals). Some of this complexity is yet to be unravelled.

A Special Building Process: the Migrations of the Neural Crest

The neural crest has already been described as forming at the point where the neural groove of the very early embryo has become a tube and just separated from the ectoderm. Populations of cells in the dorsal part of the tube on each side leave the tube and begin to migrate throughout the body. These cells are induced by factors relating to both the non-neural ectoderm and the neural ectoderm itself. The cells break free from these ectodermal layers, become more like mesenchymal cells, and migrate away. The migrations are influenced by many molecules residing in the extra-cellular matrix of the body. The migrations in the trunk have been summarized earlier. In the head, these migrations are even more complex. Indeed, the entire system of the neural crest produces an astonishing array of structures.

In the trunk the neural crest cells migrate in three primary directions, a dorsolateral pathway producing pigment cells in the skin, a ventral pathway producing the cells of the sympathetico-adrenal system, and, between these two, a ventro-lateral pathway producing the sensory ganglia of the spinal nerves.

In the head, in contrast, the neural crest is a much greater component of the cranial (cephalic) end of the embryo and forms a much wider array of tissues. These

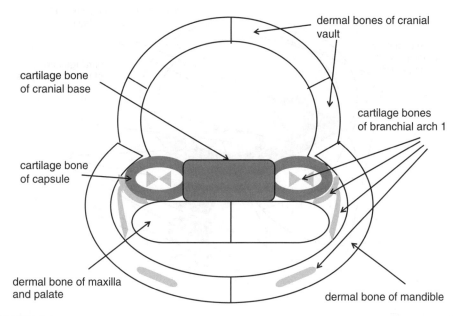

FIGURE 6.9

Diagrammatic view of a cross-section of the human showing the embryonic components that give rise to eventual skeletal elements of the skull.

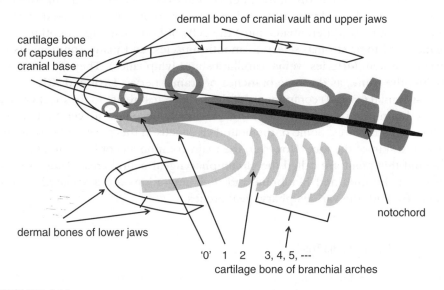

FIGURE 6.10

Diagrammatic longitudinal section of the head of a creature (such as a fish) in which the folding found in humans does not occur.

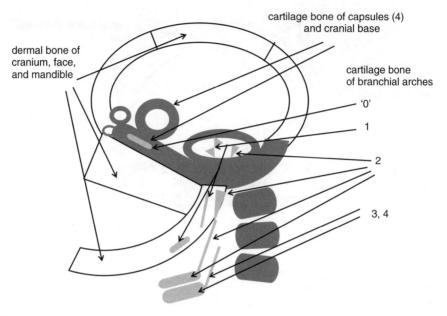

FIGURE 6.11

Same diagrammatic section in a human showing the similarities and differences with Fig. 6.10.

tissues come to constitute many of the internal tissues of the head. Thus, in addition to the cell types produced by trunk neural crest and also present in the head, cranial neural crest produces (differentiates into) the cells of bone, cartilage, dentine, dermis, glands and their stroma, selected smooth muscles, and many components of the vascular system (arteries, veins, capillaries and lymphatics). They also migrate distally into the trunk, as do some branches of cranial nerves **IXth** and **Xth**.

The most cranial (rostral) components of the neural crest relate to the most cranial part of the brain, the prosencephalon, that eventually becomes the cerebral cortex and associated structures. Some of the neural crest cells in this position become the dermis of the skin of the face. However, the main components of the cranial neural crest behind this point are called the circumpharyngeal neural crest. They arise in the region of the mid- and hindbrain and their migrations are associated, as the name implies, with the various pharyngeal arches.

6.3 Head Similarities to the Trunk

A critical part of understanding the head, therefore, is the further understanding of similarities to the trunk. This involves recognizing the relationships between dermatomes, myotomes and sclerotomes in the trunk, and their head equivalents. Some of this has already been described but it is useful to draw those descriptions together here.

6.3.1 Dermatomes, Skin, and Sensation

There are head equivalents of the trunk dermatomes. The main one comprises most of the dermal covering of most of the ventral aspect of the head (the face). It is supplied by a single nerve (on each side), the **Vth** cranial nerve or trigeminal. This is primarily the nerve of the first pharyngeal arch but it also contains the general somatic afferent (sensory) fibers from the head dermis. These fibers have their cell bodies in that part of the trigeminal ganglion that is the equivalent of a dorsal root ganglion, in other words, like the dorsal root of a spinal nerve. However, the trigeminal nerve also carries other fibers associated with the branchial structures that are not found in a typical spinal dorsal root.

Likewise, other much smaller general somatic sensory areas also exist in the head, and they are supplied by nerve fibers carried to other general somatic afferent nuclei in the brain stem via appropriate ganglia (again equivalent to dorsal root ganglia in the trunk) in other cranial nerves. These are branchial in origin, hence are supplied by specific branches of the nerves of the branchial arches. These include general sensory areas of: the **Vth** cranial nerve (the first branchial arch) that supply aspects of the middle ear and pharyngo-tympanic tube, sensory areas of the **VIIth** cranial nerve (the second branchial arch) that supply aspects of the middle ear, and the **IXth** and **Xth** cranial nerves also with minor branchial arch areas. Though small, these nerve branches are evidence, yet again, of the patterns that can be seen both from development and evolution. They also have clinical importance in relation to medical conditions of these nerves and their sensory areas.

6.3.2 Myotomes, Muscle, and Movement

The myotomal equivalents in the head are those developed from the head paraxial mesoderm in relation to brain segments: that is, from midbrain and hindbrain somit-omeres (see Chapter 7). The first three of these myotomal elements are ventral to the prechordal plate and the last two or three ventral to the hindbrain. The muscles that arise from them are clearly of spinal general somatic myotome equivalence. It is just that the somitomeres in the trunk are transient and fuse to give somites, whereas the somitomeres in the head, as we have seen, largely remain separate.

The cell bodies of their innervating nerve fibers are located in the ventral-most portion of the basal plate (see Chapter 7) just like the equivalent nuclei for the ventral (motor) roots of the spinal cord. The axons generally exit from the brain in the ventral-most locations. The specific cranial nerves are the **IIIrd, IVth and VIth** cranial nerves (for the **extrinsic eye muscles**) and the **XIIth** cranial nerve, a combination of two or three ventral roots (for the **intrinsic tongue muscles**).

The extrinsic muscles of the eye move the whole eyeball within the orbit, direct-ing its gaze at external objects of interest. These movements are obviously very complexly related; interferences in their functions can produce complex optical symptoms, like double vision. The three ventral somitomeres that give rise to these muscles are sometimes called pro-otic somitomeres because they lie cranially to the huge otic structures. There may even be more than three such somitomeres but this matter is still under investigation.

The more caudal (occipital) group of ventral somitomeres give rise to the intrinsic muscles of the tongue. Though deriving from ventral somitomeres like the above-mentioned, these somitomeres are linked with the somites of the cranial-most part of the trunk (the cervical region). As a result, these somitomeres join, becoming very somite-like and, as mentioned elsewhere, they are often called occipital somites. They migrate very ventrally and give rise to the intrinsic muscle of the tongue. There is no evidence in tongue muscle structure that it comprises several segments. However, it is supplied by a series of ventral nerve rootlets that seem to be in series with the ventral roots of the first few spinal nerves in the trunk. Instead of each forming separate ventral roots, they combine to form the twelfth (**XIIth**) cranial nerve (the hypoglossal). The propinquity and similarity of these rootlets to the motor roots of each of the first few cervical spinal nerves means that they are sometimes described as a separate group of hypobranchial nerves (and their muscles). The muscles produce changes in the position of the tongue (moving it up, down, left, and right), and in the form of the tongue (flattening it, balling it up, curling it, folding it). All of these tongue movements are important in sucking, chewing, swallowing, breathing, talking, and so on.

6.3.3 Sclerotomes, Bones, and Support

In terms of sclerotomes, trunk equivalences in the head become even more complex. The skeleton of the head indeed comprises some portions that are the equivalent of the axial skeleton in the trunk. These include the central parts of the sphenoid and occipital bones: (i) the bone from the dorsum sellae leading caudally to the foramen magnum, (ii) the bone surrounding the foramen magnum including the margins of the foramen proper, the occipital condyles (surfaces for articulation with the first cervical vertebra), and (iii) the most caudal portion of the occipital squama behind and above the foramen magnum. They are all the equivalents of the sclerotomes for the fused occipital somites and, like them, appear first as cartilaginous structures which are endochondrally converted to bone.

Indeed, the bone around the foramen magnum actually looks rather like the first cervical vertebra: a ring of bone surrounding the brain stem leading to the spinal cord. With the basi-sphenoid and basi-occiput it is probably actually two or three fused bodies of occipital 'vertebrae'. Part of the basi-occiput even looks like the body of, say, a typical cervical vertebra. This is, indeed, a correct serial homology. Like the trunk skeleton, these bones, initially evident as mesenchymal condensations, become converted into cartilage and, in the process called endochondral ossification, the cartilage is then replaced by bone.

All this is summarized in Figs. 6.8 to 6.11.

6.4   Head Differences from the Trunk

6.4.1   Skin Areas in the Head Different from Those in the Trunk

Although we have recognized some similarities in skin areas and their nerve supplies in the trunk and head, it is useful to recognize their major differences. Though the skin areas on the head are general somatic sensory, they involve structures that are of branchial origin. In addition, though the branchial structures have organs in their epithelial surfaces that are supplied by special visceral nerves, nevertheless, the branchial nerves (in addition to the trigeminal) have small components that cover general somatic sensations. The separation between these is not obvious but has been already explicated.

6.4.2   Muscles in the Head Not Represented in the Trunk

A number of head muscles move the various elements of the pharyngeal (branchial) arches. Their trunk equivalents are the intrinsic muscles of the gut tube. In the trunk these muscles are, of course, smooth muscles, not under somatic control, but visceral. The nerves supplying them are given the term 'general visceral efferents'. In the head these muscles are not smooth muscle but striated muscle, with large components of voluntary control. As a result, though they are the developmental equivalents of general visceral components in the trunk, their voluntary and conscious functions are especially dignified by the additional term 'special visceral efferents'.

Thus those muscles that are associated with the pharyngeal (branchial) arches are supplied by the special motor parts of each of the pharyngeal (branchial) arch nerves. In evolutionary terms, there is a great affinity between the pharyngeal (branchial) muscles of a particular arch and its cranial nerve. Consequently by tracking cranial nerves and pharyngeal (branchial) arch muscles and other structures we can more readily understand their homologies.

Muscles of Mastication, Mandible Moving Muscles:

Thus, the first, mandibular, arch holds, what become in humans, the ventral jaw moving muscles: the muscles of mastication, together with small dorsal muscles associated with the dorsal ends of that arch which are involved in ear and internal pharyngeal structures: some palatal and ear ossicle moving muscles. The nerves supplying these structures are therefore the appropriate parts of the nerve of the first branchial arch, the trigeminal (**Vth**) cranial nerve.

Muscles of Facial Expression

The second branchial arch, the hyoid arch, holds, what become in humans, the muscles of facial expression, together with some small but important ventral muscles involved in hyoid movements, and a small dorsal muscle involved in moderation of movements of the middle ear ossicles. These are therefore supplied by the appropriate parts of the nerve of the second branchial arch, the facial (**VIIth**) cranial nerve.

These two arches and their derivatives have evolved in humans from structures that in fishes, amphibians, and reptiles are associated with the long and complex story of the evolution from gills to jaws.

Muscles of Swallowing, Breathing, and Speaking

The remaining pharyngeal (branchial) arches (gill structures in fishes) have evolved in humans into structures involved with pharyngeal, laryngeal, and neck structures and the functions of swallowing, breathing, and speaking. The nerves supplying these structures are, therefore, the appropriate parts of the nerves of the third and later branchial arches: the **IXth** and **Xth**, the glossopharyngeal and vagus.

Some Branchiomeric Muscles of the Neck

Finally, there are some muscles that are partially of this series but which in humans are related to the neck and upper limb. These are the totally striated voluntary muscles: trapezius complex (with the lovely older name of cucullaris, the muscles of the two sides together being said to resemble a monk's cowl) and the sterno-cleido-mastoid complex. These caudal ends of these muscles have migrated into the neck and upper limb but carry part of their head nerve supply with them in branches of the **XIth** cranial nerve, the accessory nerve.

6.4.3 Bones in the Head Not Represented in the Trunk

Though some of the bones of the head (e.g. the basioccipital and basi sphenoid bones) are basically like individual bones of the vertebral column, many other bones in the head are not found in the trunk. These relate to the special developments of the head: the sensory capsules, the branchial arches, and special neural crest migrations. There are also extensive dermal bones in the head that are represented in the human trunk only by a small part of the clavicle (though much more extensively in other vertebrates).

Bone Related to Sensory Capsules

The sclerotomal bones already described are produced by a mesenchymal–cartilaginous–bony (endochondral ossification) developmental pathway. Several others bones also follow this developmental pathway, but they arise differently in relation to the main special sensory capsules, structures of the skull that are not present in the trunk.

Thus, the developing olfactory sense organs are surrounded by membranous olfactory capsules that eventually produce, together with the medially located prechordal cartilage, basal skull bone by this same process of endochondral ossification. In the adult, they can be identified as the ethmoid bones at the cranial end of the base of the skull.

The developing visual sense organs are surrounded by membranous optic capsules (together with a small cartilage, the hypophyseal cartilage) that give rise endochondrally, to rather small elements that in the adult become the roots of the greater wings, the lesser wings, and parts of the body of the sphenoid bone.

The developing stato-acoustic sense organs in the otic capsules are surrounded by membranes that, together with the parachordal cartilage, give rise endochondrally to the major portion of the petrous part of the temporal bones.

Bone Related to Branchial Arches

There is another set of skeletal elements that derive from the pharyngeal (branchial) arches. First appearing as mesenchyme in these arches, they eventually are invaded by cells (from the ubiquitous neural crest) and these become cartilage. Some of these disappear and some are replaced endochondrally by bone.

The first of these, relating to the first, mandibular arch, is called the first arch cartilage (often called Meckel's cartilage). Very little of this remains in the adult; perhaps some small ventral portion is replaced by bone in the bony mandible. Perhaps, too, a small dorsal portion has been replaced endochondrally by bone in the malleus and incus bones in the middle ear. A non-ossified structure, the spheno-mandibular ligament exists in the middle of the arch.

Similarly, there is cartilage relating to the second arch (Reichert's cartilage). Again, this largely disappears, but perhaps some small more ventral portion is converted into bone and is present in the adult as part of the body and lesser cornua of the hyoid bone. Perhaps, too, some lies in the styloid processes on the base of the skull, and in the non-osseous stylo-hyoid ligament between the two. Yet again, perhaps a small dorsal portion is present in the third middle ear ossicle, the stapes.

Mesenchymal blocks in the more caudal arches develop cartilages only at their ventral ends. These do not normally become bone in humans, but may be partially represented in the adult in some of the cartilages of the larynx.

Finally, some evidence implies that there was originally a 'zeroth' arch even more cranial than the first mandibular arch. It is possible that the skeletal element of this is represented in part by the parachordal cartilage that becomes subsumed into the very cranial end of the developing cranial base and is converted endochondrally into bone. If so, this may be a branchial arch that has long since been modified in human evolution, but that nevertheless still has its '0th' cranial nerve (see elsewhere).

Bone Related to Neural Crest Migrations

Other skull bones derive in a yet different way. They commence with populations of cells that stream ventro-laterally from the neural crest. These cell populations (of neurectodermal origin) lie peripherally in the dermis and give rise to what is called dermal bone. Dermal bones are not preformed in cartilage (as are the bones described above) but ossify ('in membrane') directly in the peripheral mesoderm (in the superficial fascia) under the ectoderm and superficial to the deep fascia.

Of course, dermal bone can occur anywhere in the body and may be an extensive part of the trunk skeleton (called the dermal skeleton) in some animals. In humans, however, the trunk components of dermal bone are few and small (perhaps little more than a portion of the clavicle, see earlier). The head components are, however, very large and comprise all of the bones of the cranial vault, both top and sides, and most of the bones of the face, and upper and lower jaws. Much of this bone surrounds and incorporates the small pieces of endochondral bones that derive from the pharyngeal arches.

As a result, problems in neural crest development may give rise to peculiar deletion deformities. In the external ear there may be anotia or microtia (total loss or reduction) of ear structures. Developmental problems of the midline linkages of the two sides may give rise to abnormalities of the jaw symphysis, cleft lip and cleft palate, and a variety of other 1st and 2nd arch syndromes: hypoplasia of mandible and maxilla, choanal atresia, vestibular dysfunctions, cochlear defects, ossicular defects, semi-circular canal defects, eye defects, microphthalmia, coloboma, and so on.

The skull, then, is a complex of originating elements that, together, coalesce into a much simpler structure than is evident in many other vertebrates. Again, these structures are shown diagrammatically in Figs. 6.8 to 6.11.

6.4.4 The Vascular System in the Head

The vessels of the cardiovascular system are as varied as the diverse organs that they supply. They differ from those in the trunk (the arrangement of which is related to somatic trunk segmentation) in that they are related to pharyngeal (branchial) segmentation. Yet at the same time, these vessels are continued into the trunk as a result of the caudal migrations of initially head structures (heart, lungs, esophagus, etc.) into the upper trunk.

Cranial Arteries

Thus, the progenitor of the heart eventually descends from a starting position at the cranial-most extent of the embryo, that is, cranial even to the developing brain. It gradually descends in the head fold, moving ventrally and caudally past the brain and other head structures, past the neck, and into the thorax. As it moves caudally it gives various arterial branches into each pharyngeal (branchial) arch. In fishes these are largely associated with the respiratory gills. In land forms they involve many other pharyngeal (branchial) derivatives.

In creatures with many branchial arches (e.g. lampreys) there are many such arterial arches, one for each arch. Because these arches extend beyond the head and far down the trunk, many of the arterial arches lie far caudally in the trunk. In bony fishes the numbers of gill arches and the number of related arterial branches are much reduced and are mainly confined to the head. In land vertebrates, especially in mammals, and therefore humans, though more branchial arteries existed in earlier embryos of some forms, most of this system disappears so that only five components persist in the human adult. These are portions of the first, second, third, fourth and sixth branchial arch vessels.

Though originally located in the head in the branchial arches, many later descend caudally with the caudal migrations of the many organs (e.g. thymus, thyroid, lungs, heart, diaphragm) into the neck and thorax. As a result they lie primarily in the neck and thorax though some derivatives do still lie in the head. Their terminal branches, the internal and external carotid arteries, do supply everything in the head.

Thus the maxillary, stapedial and internal carotid arteries remain in the head in association with the first, second and third arches, respectively. Parts of the aortic arch itself and the subclavian arteries to the upper limbs (trunk arteries) are remnants of the fourth arch arteries. Parts of the arteries carrying deoxygenated blood to the lungs (pulmonary arteries) and an artery in the fetus (that closes in the new born – the ductus arteriosus) are remnants of the sixth arch.

These arterial features are shown diagrammatically in Figs. 6.12–6.14.

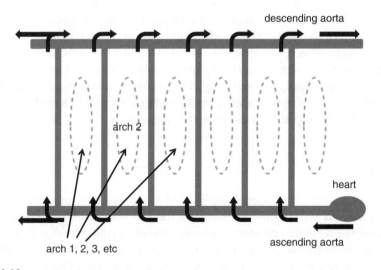

FIGURE 6.12

A plan of the theoretical pattern of arteries on one side of the head of a generalized early vertebrate (not human). The heart, bottom right, passes a major vessel (one on each side) cranially. As these vessels pass each branchial arch they give off branches passing dorsally through the arch. These vessels then coalesce into a pair of vessels (again, one on each side) that pass caudally back into the trunk. The arrows show the direction of the blood flow. The cranial-most branches give vessels that pass cranial to supply the head.

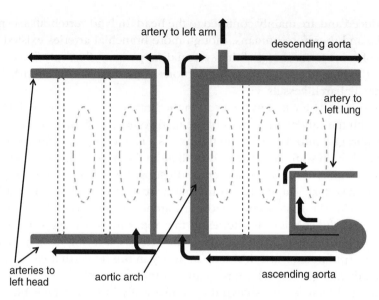

artery to left arm

descending aorta

artery to left lung

arteries to left head

aortic arch

ascending aorta

FIGURE 6.13

A plan of the actual pattern of arteries on the left side of the animal. The artery coming from the heart is the ascending aorta. Only three of the branchial arch arteries exist. The largest one, representing branchial arch 4 (and shown very wide) is thus the arch of the aorta. The caudal continuation of it back into the trunk is the descending aorta. The vessel near the dorsal aspect of the arch passing dorsalward is the artery to the left upper limb. The small sixth arch artery forms the pulmonary trunk that goes to supply the left lung. The cranial prolongation of the third arch artery gives branches (ventral and dorsal) passing forwards to supply the head, the carotid arteries. Again the arrows show the directions of blood flow.

Cranial Veins

The major veins are even more complex and variable than the arteries. Moreover, just as the arterial system embodies mostly the left-sided arch remnants (the aortic arch), so the main venous drainage embodies mostly a right-sided arch (the superior vena cava). The branchial origins of most of the smaller veins are less obviously able to be homologized with original individual branchial arches. The tributaries of this arched venous system drain blood from the head and upper limb.

Cranial Lymphatics

The lymphatic system in the head is equally a partner of the circulatory system in the head as described already in the trunk. It comprises the same components: lymphatic capillaries, lymphatic tissues, lymph nodes and lymphatic collecting vessels. These structures are present in the head as lymphatic capillaries leading

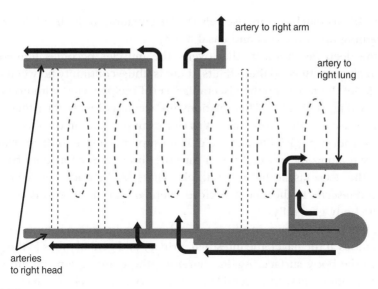

artery to right arm

artery to
right lung

arteries
to right head

FIGURE 6.14

A plan of the actual pattern of arteries on the right side of the animal. The artery coming from the heart is the same single ascending aorta as in Fig. 6.13. Again only three of the branchial arch arteries exist. However, that vessel representing branchial arch 4 (shown rather narrow) is the dorsal ward artery passing to the right upper limb. There is no descending aorta on this side. The small sixth arch artery forms the pulmonary trunk that goes to supply the right lung. As in Fig. 6.13, the cranial prolongation of the third arch artery gives branches (ventral and dorsal) passing forwards to supply the head, the carotid arteries. Again the arrows show the directions of blood flow.

from diffusely distributed lymphatic tissues (lymphoid tissue), through distinctly distributed patches (lymphoid patches), as encapsulated lymphoid tissue (lymph nodes, these last being generally confined to mammals), and as lymphatic collecting ducts (major lymph vessels).

Superficial/Deep organization of the Vasculature

These various vascular elements (arteries, veins and lymphatics) are also, as in the trunk, organized in relation to the fundamental separation of tissues into superficial and deep fascial systems. Thus there are partially separate arterial, venous and lymphatic systems within the superficial fascia and beneath the deep fascia, respectively. The separations are, of course, not total; there are connections between them; but these tend to be rather few and occupy particular sites. In the limbs, for instance, the major connections between these systems are close to the flexures of the main joints (elbow and knee, axilla and groin, respectively), an arrangement that is important in clinical problems (such as varicose veins). The situation in the head differs because of the opening up of the oral and nasal tubes.

Thus, the deep fascia sheets that overly the muscles becomes limited by their cranial attachment. Beyond these limits, there is, thus, continuity between the superficial fascia (of the face) and the visceral fascia (of the pharynx) with no intervening deep fascia (Fig. 6.15(a,b)). This relationship accounts for the fact that infections, tumor metastases, or even foreign bodies (for example) can, in the face, much more easily pass from superficial to deep (e.g. from superficial face to deep pharynx, and thence down along the visceral fascia into the thorax). Such a route bypasses the barrier of the deep fascia that is found elsewhere in the body between superficial and visceral fasciae. In the face, therefore, there are far more serious consequences than in the body generally.

This situation also occurs in the trunk, where, again, internal tubes, such as the umbilical tubes at the navel, the intestinal, urinary and genital tubes in the perineum, come onto the body surface. Again, therefore, these are regions where superficial fasciae are continuous with visceral fasciae, and where, therefore, there is continuity of arteries, veins and lymphatics that bypass the partial barrier of the deep fascia. This has similar implications for spread of infection, or tumors, or movements of foreign bodies, and so on, as in the head (see again Fig. 6.15(a,b)).

6.5 Final Head Anatomy in the Resultant Adult

The aforementioned developmental materials and the processes that change them result in the structures of the human head as we know them in the adult. These structures comprise: the cranial components of the digestive system, the respiratory system, the system of ductless glands, and the various nerves, arteries, veins and lymphatics that supply them. The brain is dealt with in Chapter 7.

6.5.1 The Digestive System in the Head

As we have seen, in the trunk, the simple endodermal tube that originally appeared in the early embryo comes to form the gut system and its glands. The regionalization of the trunk gut is established early in embryonic life via localized Hox gene expression in both endodermal (lining) layers and mesodermal (supporting) layers.

In the head, in contrast, the cranial end of the endoderm of the gut interacts with the head ectoderm to form an oral membrane. This, a double structure, gives rise to the cranial-most portion of the gut tube and is modified by the development of the pharyngeal (branchial) arches. This results in ventral oral and dorsal nasal passages leading back to dorsal pharyngeal and ventral laryngeal tubes and thence downwards to ventral trachea and dorsal esophagus in the neck, together with a number of ducted glands that evaginated from the gut endoderm and are supported by mesoderm.

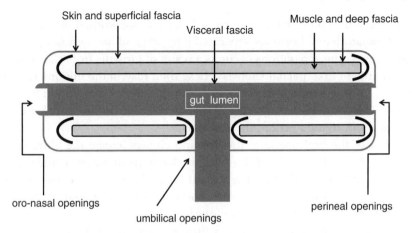

The black curves indicate pathways between superficial and visceral fasciae
in face, umbilicus and perineum

(a)

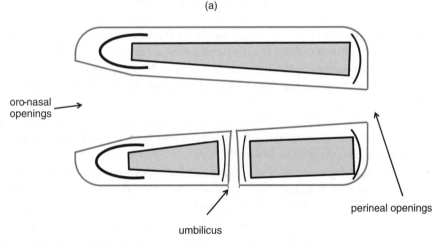

The thickness of the black curves indicates strength of pathways
between superficial and visceral fasciae in face, umbilicus and perineum

(b)

FIGURE 6.15

(a) Diagrammatic longitudinal section of a developing human (without incorporating the head and body folds that occur). The barrier provided by the muscles and bones enclosed by deep fascia can be seen to be breached in the face, at the umbilicus and in the perineum where superficial and visceral fascia come together. (b) The same diagrammatic cross-section in an adult human. The relationship between superficial and deep fascial sheets is very large in the face, considerably less in the perineum, and almost vanishingly small at the umbilicus.

Buccal Cavity and Pharynx

Thus, the originally single opening derived from the breakdown of the oral membrane (the stomatodeum) becomes separated into oral and nasal entrances of buccal and nasal cavities. The separations occur as a result of growths from the lateral walls inwards forming, at first, palatal shelves. These then fuse in the midline to form the palate proper, thus separating the stomatodeum into upper (nasal) and lower (buccal) cavities. This structure, the palate, is mainly bone but at its caudal extremity it continues as a soft tissue portion, the soft palate.

The margins of the oral opening, the lips and cheeks, create the entrance of the buccal cavity, the mouth. The lips follow the medial parts of the jaws, and the cheeks enclose the mouth more laterally. The caudal limit of the buccal cavity is a fold of mucous membrane at the caudal end of the soft palate, the palatoglossal arch, where the cavity passes more caudally into the pharynx.

The nasal cavity becomes divided by a further vertical partition (the bony part of which is called the vomer) that grows down from the roof of the cavity and articulates with the palate in the floor of the cavity. This results in a pair of nares.

As a result, the external openings on the face are the lips and the nares. These form from derivatives of the superficial fascia in relation to the dermatomes of the trigeminal nerve. Thus a central fronto-nasal process (related to the first division of the trigeminal nerve, the ophthalmic nerve) forms the deep part of the surrounds of the nares and deep to the upper lip. This comes to be overlapped by medially growing more superficial processes (maxillary processes) that form the upper lip (supplied by the maxillary division of the trigeminal nerve). A similar medially growing process, the mandibular process, supplied by the mandibular division of the trigeminal nerve, forms the lower lip.

Problems in development of the various components give rise to such deformities as hare-lip and cleft palate. The anatomy and clinical problems resulting from mal-developments are well described in many texts.

These developments also give rise to other structures in the head by evaginations. For instance, the nasal cavities evaginate pouches that excavate the surrounding bones, giving rise, in the child, and increasing in size in the adult, to air-containing sinuses. Thus sinuses gradually appear in the maxillary and frontal bones of the face, and in the ethmoid and sphenoid bones of the cranial base. They are lined with a muco-periosteum that is, at the various openings, contiguous with the muco-periosteum of the nasal passages, and are open therefore, as are the nasal passages, to the external atmosphere. The functions of these sinuses have been long debated. Certainly they do change the timbre of the voice (see the changes in speech sounds when they are blocked by infection). They also result in facial bones being much lighter than they would otherwise be. They are almost certainly involved in the warming of inspired air. Their sizes and complexity are greatest in some of the very largest animals, such as elephants, and here they would seem to be associated with vocal timbre and skull bone lightness.

A special midline, dorsal, outpouching from the naso-pharynx meets a ventral outpouching of the brain, the two together forming a midline hypophyseal placode. This forms the eventual pituitary gland which thus comprises two separate but functionally interlinked parts, an adenohyphysis and a neurohypophysis, as described previously. Eventually, however, this structure is separated from the nasal cavity by ingrowth of the cranial base beneath it (though a canal and even a cyst, indicating its original track, sometimes remain).

There are a large number of further outpouchings forming glands related to the beginnings of digestion in the head. Many of these are simple glands in the mucous membrane itself, secreting either serous and mucous fluids (or both) that help to lubricate and/or bind food particles. Some are, however, larger glands that are evaginated well outside the oral wall (much as the pancreas and liver come to lie outside the gut in the abdomen). Of course, being an evagination of the lining epithelium, they maintain their duct connection with the mouth, whereby their secretions are poured out into the mouth. The largest of these help in the trituration of foods and, through the action of enzymes in saliva, lead to the beginnings of digestion. They are the salivary glands and are named for their approximate position in the face (mandibular, sublingual and parotid).

Thus, although we tend to think of digestion as being the business of the stomach and intestines as described earlier, it begins, in fact, in the mouth. Lips, teeth, tongue, cheeks, salivary glands, all combine in helping render the food. These are assisted by the actions of a variety of the smaller oral glands that are evaginated from the oral mucous membranes.

Several other small glands in the face develop in the same way although they are not associated with digestion and the gut. These mainly comprise the lacrimal glands and also two other glands: Harderian and Duvernoy's.

The lacrimal glands lie at the outer corners of the eye fissure, and their channels, lacrimal ducts, at the inner corner of the eye pass ventrally into the nasal cavity. These glands release tears (of varying chemical compositions depending upon the state of the body) that clear the surface of the cornea. These fluids have chemically protective effects on the outer surface of the eyeball, lubricate the movements of the eyelids upon the eyeball, and are also involved in various emotional states. The lacrimal ducts are of special embryological interest because they arise as junctional tissues between the two parts of the face (the fronto-nasal and maxillary processes) already described. As such, they are supplied by nerves (the ophthalmic and maxillary branches of the trigeminal) from both sides of that embryological divide.

The Harderian glands release secretions that bathe the vomero-nasal organ in many animals. Their functions in humans are unclear.

Duvernoy's glands are found in many non-venomous snakes. Whether there is any equivalent in other animals or humans is unknown.

All these structures are similar to structures elsewhere in the alimentary tract in that they result from evaginations from epithelia. However, because of the absence of deep fascia in the face, because the internal tubes are open to the surface, they lie at the conjunction of visceral and superficial fascia already described. There are, thus, no intervening layers of deep fascia between superficial and visceral fasciae such as occurs elsewhere in the body. Hence infections and other problems of facial structures are far more serious here than elsewhere in the body (Fig. 6.15(a,b)).

Teeth

Teeth, unique to vertebrates, are embryonic derivatives of, again, the combination of two different layers: the epidermis and the mesenchyme of this region. Epidermal cells produce the enamel organ, and mesenchymal cells (of, originally, neural crest origin) produce the dermal papilla and, eventually, cells that produce dentine. There is a complex inductive interaction between these two sets of tissues. In humans, the developing teeth (usually after birth) erupt through the oral mucous membranes and thus extend into the buccal cavity. These teeth form the primary dentition and are replaced later by a secondary (or adult) dentition of the growing child and adult.

Human teeth are specialized for different functions. In the primary dentition there are, on each side of each jaw, two cutting teeth (incisors) lying centrally, a single eye-tooth (canine) lying more laterally, and two chewing teeth (molars). In the adult, on each side of the jaws, the anterior teeth of the primary dentition are replaced by secondary (adult) incisors and canines, and the two primary molars are replaced by two premolars and three molars. The terms, cutting, piercing, chewing, and manipulating can be used in relation to the functions of these teeth but, of course, masticatory function is actually considerably more complex.

Variations are frequent in eruption sequences, forms, and numbers of teeth. For example, many mammals have larger numbers of incisors (as many as four on each side of each jaw), larger numbers of premolars (as many as six on each side of each jaw, and larger numbers of molars, (also as many as six on each side in each jaw). Reductions in the numbers of teeth have occurred during evolutionary time in most primates and particularly in humans. Thus, though three molars on each side of each jaw is the norm in humans, occasionally we find four molars (like some gorillas). Occasionally, too, we sometimes find only two molars (the third molar being the one that is missing). Various other missing or supernumerary teeth may occur as variations in humans. Some aquatic mammals are born with their secondary dentition already in place. (Where is the primary dentition? Already erupted, shed, and lying in the placental membranes surrounding the fetus).

The form and proportions of primate teeth are so distinctive that they may be used in species identification. A variety of terms are applied to their forms. Cone is the name for the major cusps of the teeth and may be designated as cones or conids – for major cusps on upper and lower teeth, respectively. Major cusps may be given descriptive modifiers (such as the prefixes proto, para, etc.) to aid in identification. Minor cusps are indicated by the suffix 'ule' – for example, cuspule. A cingulum is an accessory ridge on the margins of a crown. Much of this terminology is quite old and arose from hypotheses as to how tooth form had evolved. The terms are still useful in providing accurate tooth descriptions, and are fully described in standard texts.

Tongue

One part of the tongue, its covering, derived from epithelium, contains various taste organs. These, together with taste organs on other oral structures (such as the palate), form a simple basis to the great complexities of olfactory sensations from the olfactory mucosa in the nose. Underlying the epithelium are the intrinsic tongue muscles capable of complex changes in shape, and involved, therefore, not only in alimentary functions: mastication, swallowing, and so on, but also, in humans, as complex modifiers of speech, and other communicative functions.

The epithelium and connective tissues of the tongue are derived from the first and third branchial arches (the second does not contribute to it) and hence the sensory nerve supply is from the first and third branchial arch nerves, that is, the trigeminal (**Vth**) and glossopharyngeal (**IXth**) nerves.

From the center of the region that will become the tongue, are cells that later become iodine-concentrating cells. These are the human equivalent, as it were, of the endostyle in the floor of the pharynx in so many cephalochordates (see figures in earlier chapters). These cells become the thyroid gland and are described in more detail below.

The muscles of the tongue derive from the occipital somites (as described previously) and hence the motor nerve to the tongue is the hypoglossal (**XIIth**) nerve.

6.5.2 The Respiratory System in the Head

The respiratory system develops as a caudal out-pouching from the ventral aspect of the pharynx. At first it is a single tube that gives rise to the larynx and, more caudally, the trachea. Though only a very short structure in the head, it becomes divided into right and left tubes at its caudal end. These are the rudiments of the eventual bronchi.

This inverse Y-shaped structure (Fig. 6.5) enlarges, descends below the head into the thorax, and the two limbs of the Y each arborize into a complex series of tubes, becoming smaller and smaller (bronchi and, further, bronchioles). At the end of the smallest tubes blind air sacs develop (alveoli). The arborization of tubes forms the gas-transporting element of the lungs; the berry-like terminal sacs form the gas exchange element of the lungs. Most of this is located, of course, in the neck and thorax.

The parts of the respiratory system in the head and neck are the larynx and trachea located ventrally, stemming from the ventral portion of the oro-pharynx. At the same time, the dorsal part of the oro-pharynx leads into the esophagus. As a result, this region undertakes a double function. It has to accommodate the passage of inspired gases between a dorsal tube (the nasal cavity) and a ventral tube the larynx and trachea, as well as the passage of foods and fluids from a ventral tube, the buccal cavity, to a dorsal tube, the pharynx and esophagus.

Of course, gases can pass into the stomach, hence the capacity for eructation, and solids and liquids can pass into the lungs, hence choking and coughing. In addition, reversals of these directional movements can also result. Expiration of gases from the lungs occurs with each breath, and vomiting of solids and liquids from the stomach occurs occasionally.

6.5.3 The Interaction of Breathing and Feeding

The complex movements of materials within this system are handled by, among other things, a series of temporary seals that can be set up. One such is the closure of the mouth (primarily by the special feature of lips) so that materials taken into the buccal cavity pass backwards into the pharynx and not forwards out of the mouth on swallowing. In certain clinical situations the mouth can also be sealed by the tongue being pressed forwards against the front teeth; this can give rise to what is called a 'rabbit dentition' the tongue pressing the anterior teeth forwards and thus producing a 'diastema' (gap) between front and back teeth. The mouth can also be 'sealed' by external objects, for example, the nipple closes the mouth during normal suckling, the thumb during simulated suckling, and so on.

The mouth cavity can also be separated into two parts by the tongue being raised against the hard palate. This is a second 'seal' thus separating the front of the mouth from the back (as in preparing to swallow a pill).

Yet a third seal at the back of the buccal cavity is provided by the contact of the soft palate dorsally and the back of the tongue ventrally, thus separating the food pathway in the mouth from the air pathway that crosses behind this seal.

A fourth seal occurs between the soft palate and the epiglottis, extending over the opening of the larynx.

All of these structures help channel materials (solids, fluids and gases) into the right passage during swallowing, breathing, talking, and so on.

These mechanisms themselves develop. Thus the fetus takes in amniotic fluid in intrauterine life – inhaling it into the lungs and swallowing it into the stomach. As food and oxygen are supplied by the mother from blood in the umbilical cord, nothing untoward occurs. Immediately at birth, however, the fluid in the lungs is ejected by the child (or sucked out by the midwife) in preparation for its first breath of air. Likewise, almost immediately after birth the child takes its first feeds via suckling from the nipple inserted into the mouth. Hence there are immediate changes that must occur.

Suckling involves taking food into the mouth from the nipple followed by swallowing. Because the nipple is blocking the open mouth, swallowing makes the milk pass backwards into the pharynx. Changes occur in swallowing during the conversion from liquid (milk) to solid (food) during weaning as suckling converts into swallowing, and speech develops. Misdevelopment of these mechanisms can give rise to a number of clinical problems. Speaking humans rely on the third oral seal to keep food and air passages separate during respiration, speaking and swallowing.

6.5.4 Ductless Glands in the Head

There are a series of other glands that develop in the head in relation to the digestive tubes but which are not of gut origin. They may be thought of as paralleling the various glands that lie in relation to the gut in the trunk that are not of gut origin (e.g. the insulin islets in the pancreas, and various carcinoid islets in the gut). They secrete their components into the blood stream. They are, therefore, without external ducts linking them to the gut cavity. In the head, they comprise thyroids and parathyroids together with previously mentioned glands such as the pituitary (see above).

The Thyroid

The thyroid gland produces, stores and releases into the blood stream hormones that regulate metabolic rate, growth and reproduction. These hormones contain iodine and, though initially discovered in mammals, are now known to be synthesized in all vertebrates. Indeed, the business of iodine capture and concentration is evident in all of the creatures named in this book – even such non-vertebrates as acorn worms. In these creatures the iodine-concentrating tissue, the endostyle, forms in the ventral aspect of the branchial cage, and it is in this location, the ventral aspect of the pharynx, that the iodine-concentrating thyroid gland develops in all vertebrates including humans (see figures in Chapter 2).

Thus, in humans, a thyroid rudiment develops in relation to the limits between the first and third branchial segments contributing to the developing tongue epithelium. This rudiment migrates caudally so that the eventual adult thyroid comes to lie in the neck as a bilobed structure on each side of the larynx. Its initial midline origin is indicated, at the junction between the two sides, by, commonly, a midline median lobe. Further, some thyroid tissue may remain located in the base of the tongue, an anomalous lingual thyroid. Yet further, the migration caudally may extend even into the thorax so that there may be a thyroid mass in the mediastinum of the thorax. Finally, this entire tract may be outlined by fascial connections related to the migratory pathway, and by small accessory masses of thyroid tissue lying anywhere along this track.

The Parathyroids

The parathyroid glands help control levels of calcium in the body. They are paired structures, developing from two of the branchial pouches on each side. Named purely because of their close association with the thyroid gland, they are bilateral cell masses that lie on the dorsal surface of, or may even be embedded within, the thyroid lobes. In this position, they are at risk in any pathological, surgical or traumatic interference to the thyroid.

The Ultimobranchial Body

This is also located in the thyroid. The cells of this structure are thought to be neural crest derivatives that migrate ventrally into the ventral ends of the branchial arches. It functions by counteracting the calcium raising action of the parathyroids on the blood by a secretion called calcitonin. In most vertebrates, this secretion is carried out by a separate organ located in the ventral aspect of the neck. In mammals a separate organ is not evident; the cells producing this effect are special follicular cells that are scattered amongst the iodine-concentrating cells of the thyroid.

The Thymus

This is a paired lymphoid structure that commences primarily in the third branchial pouch. It migrates even further caudally into the thorax. Accessory thymic tissue may occur at any point in this descending pathway. It fails to develop in the absence of neural crest tissue.

6.6 Head Structures and the Nervous System

We are now in a position to understand even more of the relationships of the structures in the adult human head. These depend enormously upon the formation of the peripheral and central nervous system. Because, however, of the great complexity of the central nervous system in the head, this is dealt with separately (see Chapter 7). At this point, therefore, let us examine the basics of the peripheral nerve system in the head.

6.6.1 Peripheral Nerve Supply in the Head

Main Brain Components

The **neural tube**, though initially developing in the head as in the trunk, rapidly becomes much more complex than in the trunk, foreshadowing the enormous brain. First recognizable as a thickening of the ectodermal neural plate overlying the cranial end of the notochord produced by molecular factors from the dorsal ectoderm, this region receives further molecular signals from the notochord and mesoderm to form three enlargements. These are known as the **forebrain** (hence **prosencephalon**) **midbrain** (**mesencephalon**) and **hindbrain** (**rhombencephalon**).

These three main parts undergo further subdivision: the cranial-most prosencephalon divides into a more dorsal **telencephalon (endbrain)** with prominent outpocketings that ultimately form the cerebral hemispheres, and a more caudal and deeper **diencephalon** (which already shows the optical vesicles, from the optic placodes, lying laterally). The **mesencephalon** remains undivided. The hindbrain, the rhombencephalon, starts to subdivide into a more cranial **metencephalon** and a more caudal **myelencephalon** and these two brain components are related to the hind part of the head and show evidences of some similarity to the spinal cord in the trunk. These five subdivisions of the early brain represent a fundamental organization that persists throughout life.

This organization is patterned, to a degree, in a manner similar to that in the spinal cord in the trunk. That is, it arises, in relation, longitudinally, to segments like trunk somites. However, these patterns differ in the head. First, the longitudinal segmentation in the brain, as for the peripheral structures in the head, shows its segmentation in relation to half segments, neuromeres, rather than full segments, somites, as in the trunk.

Thus, in the hindbrain the half segments are called rhombomeres produced by developmentally prominent molecular transcription factors. These rhombomeres are located, alternately, dorsal (sensory) and ventral (motor) along the developing hindbrain. For instance, rhombomeres R1 and R2 relate to sensory and motor components of the first (mandibular) branchial arch which is therefore supplied (largely) by the Vth cranial nerve (trigeminal). Rhombomeres R3 and R4 relate to sensory and motor components of the second branchial arch (hyoid) which is therefore supplied by the VIIth cranial nerve (facial). Rhombomeres R5 to R8 relate to sensory and motor components of the third and fourth branchial arches and the IXth and Xth cranial nerves. The correspondence between the rhombomeres of the developing hindbrain and somitomeres in the peripheral structures of the head is remarkable. In some ways it is a pity that the numbering systems have not been reorganized to recognize these and related relationships. However, changing the numbering at a time when this is all well recognized might create even further confusion in making correlations with older sources.

Patterning in the midbrain is present molecularly rather than obviously structurally, but the equivalent molecular components may be called mesomeres (M; two are currently recognized and dorsal M and a ventral M). The effect of this arrangement on head structures is much less obvious (or not yet fully known!).

Patterning in the forebrain is evident as a set of six prosomeres (numbered in reverse order, from P1 more caudally to P6 most cranially). These all extend from the midbrain/forebrain boundary to the tip of the forebrain. The ventral three (P1, P3 and P5) form deep ventral structures in the brain (e.g. the thalamus, a major relay station between the cerebral cortex and the body). The dorsal prosomeres (P2, P4 and P6) contribute to more dorsal brain structures (such as the cerebral hemispheres). There is even an effect upon head structures related to the brain segments (dorsal-most being related most to the special senses, and ventral-most to special motor structures but with enormously complex relationships among them.

6.6.2 Further Details of the Cranial Nerves

Though the numbering of the cranial nerves is enshrined into law (by the early human anatomists who saw many of these things first) as the **Ist** to **XIIth** cranial nerves, they are actually associated into groups in relation to the aforementioned developmental arrangements.

The Cranial 'Nerves' of the 'Special Senses'

One group of cranial nerves is associated with the special paired dorsal sensory placodes and therefore (as already described) with smell, sight, and hearing/balance senses. Though dorsal in position (sensation is located generally dorsally within the nervous system) these nerves are not the analogs of the dorsal nerve roots of the spinal cord. They are additional 'ultra-dorsal' special sensory elements that occur only in the head (but see the lateral line system in fishes, Chapter 3).

The olfactory and visual primary sensory nerves are very small nerves that are located within the peripheral sensory structure.

In olfaction, the sense organs are the sensory epithelial cells in the nasal mucosa. The short primary afferent cells pass from these sensory organs through numerous small passages in the cranial base (at that point called the cribriform plate of the ethmoid) to the closely adjacent olfactory bulb. Apparently lying outside the brain, this is actually a brain structure that is connected to the brain by a long tract called the olfactory tract. In earlier days this tract was described as the first cranial nerve. However, it is now known to be a brain tract and the true olfactory nerves the aforementioned very short primary afferents.

Similarly, for vision, the sense organs are rod and cone cells lying in the retina; the true afferent neurones are very short cells, also in the retina. These synapse with second order cells that commence in the retina but that pass into the optic 'nerve', backwards to the optic chiasma and backwards still into the brain. Again, in earlier days, the post optic portion of this tract was described as the second cranial nerve. Again, however, though apparently lying outside the brain, these components (optic 'nerve', optic chiasma and optic tract) are now recognized as a single brain structure conveying visual information from the primary optic nerve cells backwards to the brain.

Thus these olfactory and optic brain structures, though lying very close to their primary sense organs, are extensions of the central brain, not peripheral nerves. Associated with this is the fact that, like the rest of the brain, they have the same coverings (dura and pia mater). However, the term 'nerve' for these particular external tracts is hallowed by use and many texts still refer to them as the first and second 'cranial nerves'.

This means, however, that the subdural and subarachnoid spaces, normally surrounding the brain and buried deep within the cranium, actually reach out towards the external world through the coverings of these external tracts. These coverings, and the space that they contain, with its cerebro-spinal fluid, are within a millimetre or so of the external world.

As a result, a small child's exploration of its own nasal cavity with (say) a pencil, or a fracture of the ethmoid during a fracture of the anterior cranial fossa may result in passage of cerebrospinal fluid down the nose, and therefore, also, the ready passage of microorganisms from a traditional dirty region back up to the brain. Likewise, damage to or infection in the eye therefore also means that resultant infective organisms are perilously close to the brain coverings of the optic nerve and its contained cerebrospinal fluid.

The situation is not quite the same for the vestibule/cochlear nerve involved with hearing and balance. Here, the sensory organs are cell bodies in the spiral organ of the cochlea (for hearing) and cell bodies in the vestibular apparatus (for position and movement). The primary afferent nerve fibers (the vestibular/cochlear nerve) pass to the brain. Yet this is a very short nerve, and though it seems deeply protected within the skull, a fracture of the middle cranial fossa, or a foreign body driven through the external auditory meatus, can result in a connection with the external world, with cerebrospinal fluid dribbling out of the ear, and organisms from the external world passing in the reverse direction to the brain.

The Cranial Nerves of General Somatic Sensation

A second group of cranial nerves is truly not unlike the dorsal roots of the spinal nerves. They arise, like the sensory roots in the spinal column, from the dorso-lateral parts of the internal brain structure (the sensory columns). They provide general sensation to the head, through the sensory portion of the **Vth** cranial nerve. They differ, however, from their trunk analogs, partly in not joining up with the equivalent ventral nerves (**IIIrd**, **IVth**, **VIth**, and **XIIth**) and partly in being organized mainly through what seems to be a single cranial nerve, the **Vth**. This nerve does, however, have three major sensory branches and these provide almost the whole sensory innervation of the frontal and lateral aspects of the head and neck. The **Vth** nerve also includes other elements, both branchiomotor and special viscero-sensory relating to branchial arch derivatives. In other words, the **Vth** cranial nerve is a complex combination of nervous elements.

The Cranial Nerves of General Somatic Movement

A third group of cranial nerves are truly like the ventral roots of the spinal nerves. That is to say, they arise from the ventral-most parts of the internal brain (the ventral motor column) and supply regular striated muscles (the muscles moving the eyeball and the muscles of the tongue). These are the equivalents of the striated muscles of the trunk. Again, like the general somatic sensory nerves, they differ from their trunk analogs in not joining up with equivalent dorsal roots. They remain separate as the cranial nerves numbers **III**, **IV**, **VI** and **XII**. In this relationship, that is, in not joining with the sensory roots to form the equivalent of a spinal nerve, they reflect a much earlier arrangement evident in very simple vertebrate relatives.

Cranial Branchial Nerves, Special Visceral Efferent and Afferent Components

A fourth and fifth set of nerve fibers arise from the brain in positions intermediate between the ventral (somato-motor) column and the dorsal (somato-sensory) column. In this sense they are similar to the effector set of fibers like the efferent viscero-motor and afferent viscero-sensory sets of fibers that, in the trunk, lie between the ventral and dorsal columns of the spinal cord. But in the head these are much more complex in relation to pharyngeal (branchial) segmented structures.

Thus the effector (instructions to do something) sets of such fibers are elaborated into general viscero-motor (non-segmented) autonomic motor fibers to the smooth muscles and glands of the head, as in the trunk, and into special viscero-motor (or branchio-motor) fibers to the striated muscles of each of the branchial arches. The branchiomotor components are large and form the major parts of these nerves.

The branchiomotor components of the **Vth** nerve (trigeminal) serve the striated muscles of the first arch, largely the muscles of mastication plus some small muscles associated with the middle ear.

The branchiomotor components of the **VIIth** nerve (facial) are motor to the striated muscles of the second arch, largely the muscles of facial expression plus other small muscles that, like those of the first arch, are associated with the middle ear.

The branchiomotor components of the **IXth** and **Xth** cranial nerves (glossopharyngeal and vagus) are motor to the striated muscles of the third and fourth arches, largely to the muscles of the pharynx and larynx.

The afferent nerve fibers (carrying information from structures, or afferents) in these nerves likewise evolve both into fibers conveying general visceral information from tissues, vessels and glands of the head (as in the trunk), and special visceral afferent fibers (from the special sense organs, e.g. taste) of each of the branchial arch. Again, these special branchial sensory components are the much larger portion of this afferent suite of fibers.

They include in the first arch, the special visceral sensory fibers of the **Vth** nerve (the trigeminal) subserving taste on the anterior two thirds of the tongue. In the third and fourth arches they include special sensory components of the **IXth** and **Xth** cranial nerves (glossopharyngeal and vagus), largely taste in the posterior part of the tongue and probably in some other oral and pharyngeal mucous membranes.

Cranial Autonomics

The general visceral fibers in these cranial nerves are associated with involuntary efferent (motor) activities (e.g. contractions of smooth muscle and secretion of glands) and unconscious afferent (sensory) information (about the physiological states of internal organs). A large portion of them extends caudally into the trunk to participate in very important trunk functions, as a large outflow into the trunk through extensions of the **Xth** cranial nerve, the vagus, extending as far distally as the derivatives of the mid-gut in the abdomen.

(It should also be noted that some trunk visceral nerve components, from the sympathetic chain in the trunk, also supply sympathetic structures in the head through reverse outflows from the cervical sympathetic chain into the head).

6.6.3 Sense Organs, Sensation and Sensory Nerves in General

To survive, organisms must react to danger and take advantage of opportunity. This requires, amongst other things, information about the external environment, knowledge of the body's internal physiology, and records of previous experience. Previous experiences from sensory receptors are recorded within the nervous system whether as conscious memory or unconscious information. Of course, the whole body contains such a system recording both internal and external states of play, as described earlier. However, the head contains special sensory organs because this is the part of the organism that goes first into the environment, that takes in food, that apprehends dangers, that recognises conspecifics and mates, and so on. Accordingly, over the eons of evolution, these cranial receptors have evolved in special ways.

Thus, some receptors are associated with sensations that we, as humans, are aware of. For example, humans perceive in color, many other animals only in black and white. Yet, vertebrates differ enormously in such abilities. Many creatures perceive in rainbow hues far beyond those of humans (for example many insects 'see' into the ultraviolet, colors in flowers that we cannot see, snakes into the infrared 'seeing' heat that, again, we cannot 'feel'. Many animals (dogs, snakes) sense chemicals that we cannot smell. Hawks and snakes perceive movement that we do not recognize. Bats, dogs and elephants recognize sound at frequencies lower than we hear. Some animals sense information (e.g. about the earth's magnetism) which (as far as we know) we do not recognize at all. Many animals have special mechanisms for knowing mates, enemies, or emotional states that are very little developed in humans (though perhaps more than we think).

Many sensations arise from the organism's own subjective interpretations. The environment contains the chemicals, photons, forces, and electromagnetic radiations that are 'sensed'. But the senses of taste, color, position and movement and, where available in particular species, position in the larger environment, are brain interpretations of these external phenomena.

Similarly, pain does not exist in the environment. We cannot measure pain as we can temperature, or force, or solar radiation. Pain (as taste and color) is a perception arising from events within the brain itself. We should be clear, therefore, to distinguish the various environmental stimuli from our interpretations of the stimuli. We speak of the sense of smelling, seeing, hearing, tasting, and so on as if the stimulus and the perception of the stimulus were the same. We should, however, always remember that they are not.

Sensation Within the Trunk

Of course, there are sensory organs throughout the body. Some are very simple, free nerve endings in the tissues that respond to stimuli that produce perceptions of pain, or temperature. Some of these result from external environmental stimuli, but many from internal phenomena, such as swelling and stretching, pressure, and so on. Many perceptions of pain result from perceptions of these other phenomena, such as pressure and temperature, when these are carried to excessive lengths.

Others are rather more complex forming specialized structures that are encapsulated: that is, sensory nerve fibers wrapped in mesodermal cells located in the dermis. Some of these, Meissner's corpuscles, respond to touch, others, Ruffini's and Krause's organs, to heat and cold, yet others, Pacinian corpuscles, to pressure. Such encapsulated sensory organs operate through deformation of nerve endings located within the capsule. All convert the stimulus into electrical phenomena in the nerve ending.

A special set of these sensory receptors is associated with information that is called proprioception. This is information from receptors located in muscles, tendons, joints, and fascial sheets. These receptors monitor the state of contraction of muscles, the positions of joints, the stretching of fascial sheets, ligaments and tendons, and, hence, the positions and movements of bodily parts. Their information content is particularly provided in comparison with the prior state of those receptors. Adaptation to such stimuli, to information from such receptors, readily occurs.

The information passing to the nervous system is readily associated with the spinal nervous system's response, a response that is called a reflex. Such information is also passed to higher levels of the nervous system whereby it may modulate muscular activity more widely than just a reflex (e.g. in the cerebellum) or even so far cranially that it enters into the cortex and into consciousness.

6.6.4 Sense Organs, Sensation, and Nerves in the Head: the 'Special Senses'

Though such sensations are evident throughout the entire animal, it is not surprising that some senses arose, in evolution, very early, and that, in particular, special senses arose in relation to the development of a head as described.

More about the '0th' Special Sense: Chemo-detection, Pheromones

The **0th** sense is that of the 'true' first special sensory nerve in the head. It is therefore supplied by a 'true' first cranial nerve. The sensory information seems like a form of smell but it is mainly restricted to very special 'smells', the detection of pheromones from other conspecifics providing positive or negative information about social recognition, sex relationships, inter-individual competition, even of danger, either from others members of the species, or from other species or other externals. Long unrecognized in humans, this system does exist.

The organ in other animals is the vomero-nasal organ and it is supplied by the so-called 'nervous intermedius'. The sense associated with it may be called vomero-olfaction to differentiate it from general olfaction. In many texts the organ is spoken of as an accessory olfactory organ. Indeed, in many vertebrates it seems to be reactive to the chemical composition of food in the mouth, the chemical composition of water or air in respiratory intakes, and so on. However, it is especially sensitive to chemicals important in social, reproductive, emotional or agonistic behaviors – pheromones. Pheromones are easily seen as important in mammals in general and in non-human primates. But there is no doubt that they are also important in some humans in some social, emotional, sexual or agonistic situations.

More about the 1st Special Sense: General Chemo-detection, Smell

The sense of smell per se, olfaction, involves chemoreceptors that are mostly located in the apex of the nasal passages. These are the sensory cells of the olfactory epithelium. They communicate through their central axons with the cells of the olfactory bulb. However, although located elsewhere in the head, in the mouth and pharynx, the sense of taste is intimately associated with that of olfaction. This sense involves much cruder sensations than smell but, together, they provide a very wide range of chemical sensory information. The senses of both smell and taste are complementary. They tend also to reduce with age, hence the need for strongly spiced foods, the loss of appetite, the disinterest in food, and resultant under-nutrition that sometimes occurs with age. Yet the nerve cells of smell are amongst the few nerve cells that are very readily replaced during life.

More about the 2nd Special Sense: Electromagnetic Detection, Sight

This is the sensory organ that detects information about the intensity, wavelength and direction, of the electromagnetic spectrum. Probably no organism can tap the full range of these spectra. Many insects can detect the ultraviolet, some snakes and bats, the infrared. Further, some animals use receptors other than the eye (using specialized skin surface receptors) to tap into this spectrum. We, as humans, are stuck with what we call 'the visible spectrum' because **we** can 'see' it – between 380 and 760 nm: a very narrow range (but see the pineal gland).

A 'Wrinkle' on Sight, the Pineal Organ

In many vertebrates the roof of the central portion of the developing brain (the diencephalon) produces a single, midline photoreceptor called the parietal organ. In some forms it may actually have a kind of lens and certainly a region of photosensitive cells that lead to nerve fibers passing to various brain regions. There is great variation in this organ from group to group. There have also been many phylogenetic changes in the function of this organ though it participates in photoreception in many vertebrates.

In amniotes, however, it has also become involved in the modulation of other endocrine organs. Thus, in humans, it is associated with the release of hormones from the adenohypophysis, in heightening vasopressin effects of the neurohypophysis, in inhibition of activity of the thyroid, and even in stimulation of components of the immune system.

Also in mammals, including humans, though the gland has lost direct photosensitivity, it still seems to respond to light via a complex pathway that includes a subset of retinal ganglion cells containing the photopigment, melanopsin. The pineal also shows circadian oscillations, but these damp down within a few days in the absence of input from primary circadian pacemakers in the brain. It does seem to be involved in functions related to the cycle of dark and light seen from one day to the next, and changing from one season to another. It seems especially to be involved in changes in dark/light rhythms related to rapid travel around the earth, something not possible until humans invented aeroplanes.

The nerve supply of the pineal gland and its connections to the brain, are not clearly understood. In early life in humans the organ has a truly glandular structure which reaches its greatest development at about the seventh year. Later, especially after puberty, the glandular tissue gradually disappears and is replaced by connective tissue. Tumors of the pineal are often associated in younger individuals with a variety of sexual syndromes.

More about 3rd Special Senses: Mechano-detection, Balance, Hearing

Early in evolution, in creatures that lived in water, sense organs evolved for detecting small changes in mechanical forces on the body due to water currents, to the movements of the body in water, and to the detection of sound waves moving through water. These involved a system called the lateral line system, most concentrated in the head that goes first into the environment, but also extending along the length of the body and into the tail. This system responded to external currents in the water, to currents generated by the animal's internal movements within the water, and to the mechanical effects of sound waves transmitted to the animal through the water. It was carried out by the stimulation of hair cells through movements of fluids flowing through the lateral line canals. This system was for long thought to relate to the **VIIth**, **IXth**, and **Xth**, cranial nerves, the facial, glossopharyngeal and vagus. It is now thought to be a separate set of cranial nerves that were discovered so late that they have not been intercalated into the earlier numbering system of the other cranial nerves.

With the evolution of land living creatures, the lateral line system, devoted to the detection of water currents in these three modes, seems to have been lost. But the senses concerned are still required on land, and changes occurred to isolate the fluids in the head component of this sensory system. In mammals and humans, these organs are present as the complex of vestibular/cochlear (stato-acoustic) organs. One of these is the vestibular apparatus. This can respond to rotations of the head through hair cells that are influenced by the movements of their contained fluids. A second component is the cochlea and this responds to sound, again stimulated by sound waves within its contained fluids.

In mammals, head position involved fluid movements in the internal tubes of the semicircular canals and the sacculus and utriculus of the vestibular apparatus. This allows sensation of the animal's position in space. Likewise, in mammals, sound waves in air are received by fluids in the cochlea whereby sound is sensed. However, in the passage from air to the fluids in the internal ear, the sound waves enter via a new set of structures, the external and middle ears. These structures did not arise 'de novo'. As is usual in evolution, they were adapted from a variety of already existing organs in the first and second branchial arches. In this system – the external ear, the middle ear and its auditory ossicles – sound is transmitted to the inner ear where it is converted into pulsations in fluids in the cochlea of the inner ear. Both of these mechanical modalities are recognized through the **VIIIth** cranial nerve.

Additional Special Senses

Chemical, electromagnetic and mechanical stimuli are not the only types of information which vertebrates can sense. Sensing electrical information in the environment, and producing electrical phenomena within the body are also important in a number of creatures. For example, environmental electrical information can act as part of a navigational system and may be used in social signaling in many fishes; in some specialized fishes electrical discharges from electric organs can be produced.

Such special electrical senses seem not to be important in humans. (But then sensing pheromones were, for a long time, thought not to be important in humans.) Yet sensing electrical phenomena is evident in the reactions of cells to electrical activity, and in the adaptations of bone to mechanical stimuli. Perhaps further study is needed here.

Whether there are any other, hitherto unrecognized sensory modalities must remain a somewhat unlikely but nevertheless open possibility.

6.6.5 Special Visceral Senses Related to the Development of Branchial Arches

As summarized above, there is another set of senses that appear to be similar to olfaction. Indeed, in functional terms, they are ancillary to olfaction. Though they work with olfaction, they are more primitive and less sensitive than olfaction. They are associated with the internal functions of the gut tube. In the trunk, such 'senses' (afferent information) though evident throughout the gut are generally not apparent to our consciousness. In the head, however, they are associated with the special development of the branchial segmentation, and the afferent information to the brain that they provide is conscious. They are chemical receptors that sense chemical features of the food (taste) in relation to the intake and initial processing of food in the cranial part of the gut.

In aquatic vertebrates such chemoreceptors are not only found in the cranial gut tube but are widely distributed across the surface of the body. In land vertebrates, and especially in mammals and therefore humans, these sensory modalities are confined largely to the branchial components of the mouth. Though much less sensitive than olfaction, as discussed above, they are intimately related functionally to olfaction. However, their origin is branchial in contrast to the neural origin of olfaction.

6.6.6 Putting It All Together

The foregoing sections have pointed out the many relationships among the different head structures. They derive from complex patterns of function, development, and evolution. They are complex and so an overall summary is useful. This is provided in Figs. 6.16 and 6.17. Figure 6.16 presents the simpler relationship within the trunk and therefore summarizes earlier chapters. It can be compared with Fig. 6.17

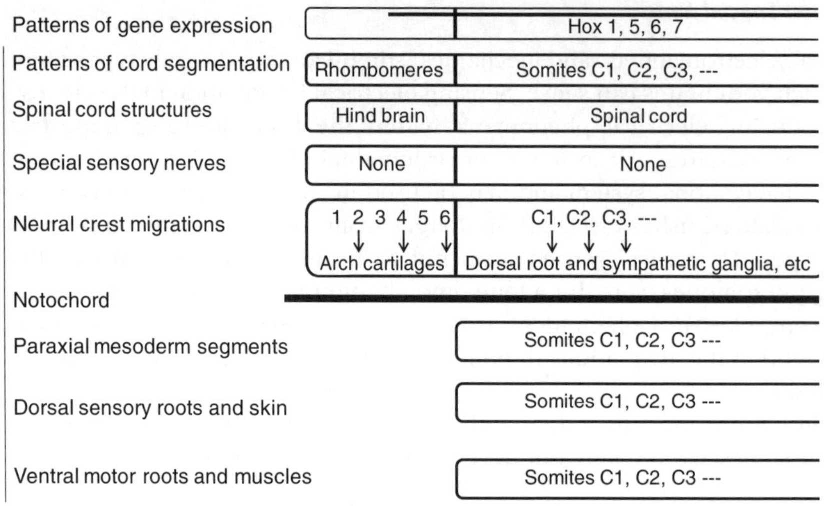

Patterns of gene expression		Hox 1, 5, 6, 7
Patterns of cord segmentation	Rhombomeres	Somites C1, C2, C3, ---
Spinal cord structures	Hind brain	Spinal cord
Special sensory nerves	None	None
Neural crest migrations	1 2 3 4 5 6 ↓ ↓ ↓ Arch cartilages	C1, C2, C3, --- ↓ ↓ ↓ Dorsal root and sympathetic ganglia, etc
Notochord		
Paraxial mesoderm segments		Somites C1, C2, C3 ---
Dorsal sensory roots and skin		Somites C1, C2, C3 ---
Ventral motor roots and muscles		Somites C1, C2, C3 ---

FIGURE 6.16

A further diagrammatic longitudinal (unfolded) view of hindhead structures, including patterns of gene expression and neural tube segmentation in relation to extra-neural structures (compare with Fig. 6.5).

showing the equivalent relationships in the head. The relationships of molecular factors, embryonic segmentation, and head components are clear. The notochord, though completely lost in the adult, is an especially important developmental marker and organizer.

6.7 Heads over the Long Haul, from Lampreys to Humans

As with the trunk, the structure of the human head can also be viewed through the lens of comparative anatomy. Some information about this has been provided in prior introductory chapters. It is useful, however, to cover this in more detail here.

The most obvious element of the head is the skull. Its arrangements in different creatures reflect separate phylogenetic sources as well as distinctive functional adaptations. The evolutionarily most ancient part is the branchial or pharyngeal apparatus (also called, in appropriate species, the splanchnocranium or visceral cranium). This is a component that exists even in those protochordates where there is no skull and precious little trunk. It is the supporting structure for their pharyngeal slits. This component also exists within the early human embryo, but largely disappears being replaced by the second part: the chondrocranium.

The second part, the chondrocranium, is formed of cartilage, or of cartilage replaced by bone, and is evident in the later craniates. It underlies and supports the

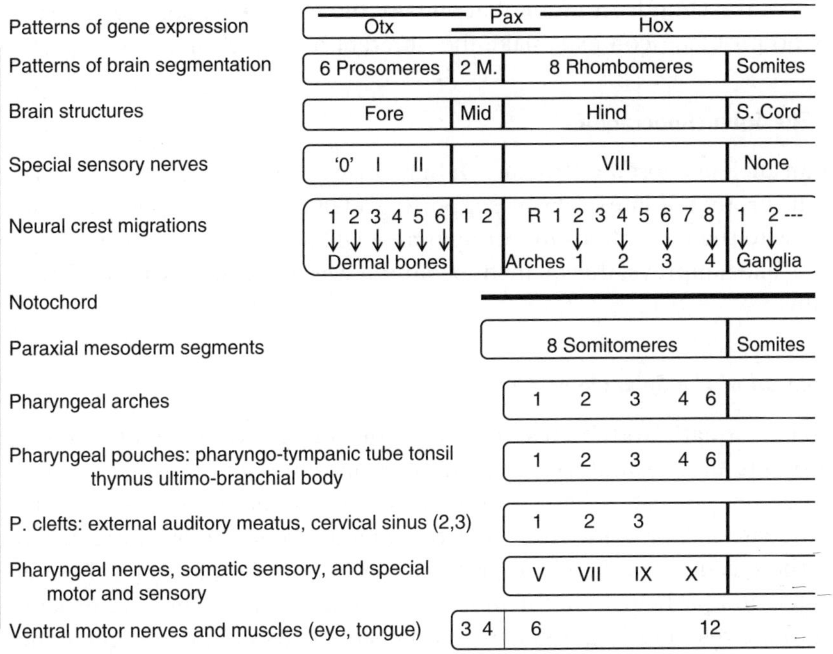

Patterns of gene expression	Otx	Pax	Hox	
Patterns of brain segmentation	6 Prosomeres	2 M.	8 Rhombomeres	Somites
Brain structures	Fore	Mid	Hind	S. Cord
Special sensory nerves	'0' I II		VIII	None
Neural crest migrations	1 2 3 4 5 6 ↓↓↓↓↓↓ Dermal bones	1 2	R 1 2 3 4 5 6 7 8 ↓ ↓ ↓ ↓ Arches 1 2 3 4	1 2 --- ↓↓ Ganglia
Notochord				
Paraxial mesoderm segments		8 Somitomeres		Somites
Pharyngeal arches		1 2 3 4 6		
Pharyngeal pouches: pharyngo-tympanic tube tonsil thymus ultimo-branchial body		1 2 3 4 6		
P. clefts: external auditory meatus, cervical sinus (2,3)		1 2 3		
Pharyngeal nerves, somatic sensory, and special motor and sensory		V VII IX X		
Ventral motor nerves and muscles (eye, tongue)	3 4	6	12	

FIGURE 6.17

A yet further diagrammatic longitudinal (unfolded) view of head structures including the rostral-most components. Patterns of gene expression, neural tube segmentation, and neural crest migrations are shown in relation to extra-neural structures (compare with Fig. 6.5).

brain, and also partially replaces elements of splanchnocranial branchial support and movement structures. A few remnants (converted into bone) may still be present in the adult human head (for example, within the mandible).

The third part is the dermatocranium, formed of dermal bone and evident in most vertebrates. The trunk dermal skeleton may be very large in many species, but in mammals these components are much reduced, and in humans they are present in little more than a part of the clavicle. In the skull, in contrast, the dermal skeleton is extensive and gives rise to the outer casing and thus roofs over the brain. In modern fishes and amphibians many of the dermal elements have been lost or fused so that the number of bones is reduced and the skull somewhat simplified. In mammals the bones of the dermatocranium predominate, coming, in the adult, to form most of the bones of the skull, indeed hiding, from external view, most of the reduced parts of the splanchnocranium and chondrocranium (some parts of which even disappear).

Other terms are also used but the above are the fundamental building blocks. Thus the brain case is a compound of the chondrocranial base and the dermatocranial vault. The neurocranium is the further compound of the brain case together with the parts of the chondrocranium that enclose the various sensory capsules. There is usually no problem; contexts make the different usages clear.

6.7.1 The Splanchnocranium

This is an ancient chordate structure. It first arose to support the slits of the branchial basket in protochordates. These supporting elements form from local mesoderm giving rise to the unjointed supporting branchial basket. Thus in the fore-runners of creatures like lampreys, the splanchnocranium was associated with the filter feeding apparatus.

The Skeleton of the Splanchnocranium

In primarily aquatic vertebrates, the splanchnocranium of the adult is the skeleton that supports the respiratory gills. These supporting structures of the gills (pharyngeal arches) in fishes have a different origin in vertebrates as compared with protochordates. They come from cells that migrate ventrally from the dorsal neural crest. (There is no neural crest in protochordates.) Further, in fishes these elements are jointed and participate in the movements of the arches. They are usually associated with the respiratory gill system and hence are also known as gill arches. They are segmented in a separate, branchial, series that is quite distinct from the segmentation, into somites, of the trunk. One or more of the most cranial arches come to be associated with the mouth, forming jaws and supporting teeth. In these forms, the jaws are complex structures comprising elements of a number of different arches.

These arch structures become less and less obvious in land vertebrates (tetrapods) being, perhaps, least obvious of all in mammals though, of course, they remain the fundamental underpinnings of their anatomies. This relates, of course, to the respiratory structures in true tetrapods being no longer gills of the pharynx, but new structures, lungs, evaginated from the pharynx. The various cartilages developing in relation to the gill structures in fishes and related forms become branchial structures present in the mammalian embryo. Later they largely disappear and become, in the adult, replaced by bone of dermal origin.

Those few skeletal elements that do remain form, for the first (mandibular) arch (using the standard numbering system) two small ossicles in the middle ear – the malleus and incus – and for the second arch, the third ossicle in the middle ear – the stapes. The remaining arches give rise to the supporting structures of the hyoid, laryngeal, and pharyngeal apparatuses. The hyoid actually becomes bone, but the remaining structures are, in humans, usually only represented in cartilage except that sometimes they become partially bone in very old individuals.

Musculature of the Splanchnocranium

Associated with these branchial segmental skeletal structures are various soft tissues: the muscles that move them and the neurovascular bundles that supply them.

In fishes these structures are involved in the movements of the jaws and gills. They are segmented in a manner parallel to that of the skeletal supporting structures.

In terrestrial tetrapods in general, and in mammals and humans more specifically, gills, of course, do not exist. But the developmental materials that gave rise to gills in the fishes give rise in these forms to related structures. Thus the jaws, which are developmentally different structures in different mammals, minimally involved with 1st arch cartilages that largely disappear being replaced with dermal bone, are moved by the true first arch muscles, the muscles of mastication. In addition, other very small muscles (but incredibly important functionally) influence the movements of the other tiny bony elements of the first arch, the malleus, and incus in the middle ear, together with yet other small muscles moving other parts of the remaining first arch structure (e.g. the pharyngotympanic tube and the palate).

Likewise, in humans, the second arch bone is the stapes, the styloid process and the cranial half of the hyoid bone. These elements are moved by muscles related to this arch. The biggest muscle block comes to include superficial muscles in the face (often called the muscles of facial expression). These are most powerfully developed in some special mammals. For example, in whales the muscles of facial expression comprise many hundreds of muscular fascicles that are responsible for changing the shape of the melon in the face (an acoustic lens) as it organizes the direction of sounds beamed out as a kind of sonar. In elephants they consist, again, of enormously complicated muscles that, among other things, produce the complex movements of the trunk. In humans these muscles, though much less complex than in the above special examples, are still complex enough to produce the very wide range of facial expressions of which humans are capable. There are also a number of very special small muscles associated with the ear (the ear ossicle known as the stapes) and the links between the stapes and the hyoid bone (the styloid process and its connective tissue – stylo-hyoid ligaments). These are also deep components of the second arch.

In the same way, there are muscles associated with the more caudal arches. These are, however, generally more compounded with one another so as to form the overall muscles of the larynx, pharynx, and upper part of the esophagus. Given the nature of the differences that exist among the various vertebrates, it is difficult to be precisely sure which of these more caudal arches relates to which structure. That the current structure in humans has evolved from such an overall pattern is not, however, in doubt.

Nerves Related to the Splanchnocranium

Associated with these skeletal arch elements are, of course, the various nerves (certain cranial nerves) that supply these arch moving muscles, and receive appropriate sensations. These are, therefore, also in series for each gill arch.

There is, even now, considerable controversy about the evolutionary origins of these various structures, but the underlying idea of branchial segmentation would appear to be correct. The numbering of the arches, the degrees to which given arches may be more complex, and the possibility that arches may have disappeared during evolution, are still matters of conjecture. They are covered in greater detail in Chapter 7.

6.7.2 The Chondrocranium

Some elements of the chondrocranium in the head appear to lie in series with the bodies of the vertebrae in the trunk.

It was this that allowed Goethe (of Dr Faustus fame – although the idea was earlier published by Oken) to propose that the skull was simply a series of fused vertebrae. Parts of this idea persisted for many years, were adopted by Owen, and formed part of the many controversies between Owen and Huxley (Darwin's Bulldog). It turns out that Goethe and Oken and others were wrong. The back of the skull, that part of the occipital and sphenoid bones forming a base and ring around the nervous system exiting the skull, probably really is a minor extension, as it were, of the trunk skeleton into the head. But the main skeletal elements of the head are not related directly to trunk vertebrae. Indeed they could not be so because, as we have seen, this part of the head is earlier in evolution. However, these early workers were almost right. A great deal of the head is related to segmentation, just not trunk segmentation. In fact there may be a reverse element here; the caudal part of the early head system may be the thing that became modified as the trunk became grafted onto the head.

Skeleton of the Chondrocranium

Thus the chondrocranium actually derives from cranial mesodermal structures. In some fishes the entire compound of chondrocranial elements becomes fused, supporting the brain from below. This chondrocranium is an expanded and enveloping structure that supports the brain from below and at the sides. Its various parts, some related to elements like somites in the trunk and some related to the sensory capsules, fuse early, and serve as cartilaginous underpinnings for the developing brain and a surrounding support for the developing special sensory capsules.

In most vertebrates, especially mammals, however, the chondrocranium is primarily an embryonic structure. Many of its parts disappear. In most mammals the chondrocranium serves as a scaffold for the later bony skeletons that come to support and enclose the developing brain and special sense organs, and forms a bony base for the attachment of the much more extensive dermatocranium.

Thus the chondrocranium is really a double structure. Originally dorsally located neural crest cells contribute to the sensory capsules (which are dorsal) and a cranial base structure called the trabecula. We already know that they also contribute to the branchial cartilages and, therefore, because the trabecula is also derived from neural crest cells, it is likely that this is also developmentally a branchial arch that has secondarily become agglomerated into the skull base. If this were so then this would be skeleton of the '0th' arch. Such a missing arch skeleton fits well with other components that exist in the head.

Mesenchyme of mesodermal origin contributes to all the other structures of the chondrocranium. These elements fuse giving rise to the midline structure called the cranial base. This thus comprises from cranial to caudal, a central structure that ossifies as a series of separate elements (ethmoid plate, sphenethmoid, and basisphenoid, linking into a more caudal head/trunk somite series: basioccipital, supraoccipital, and exoccipital elements. Collectively, then, all of these expanded and fused elements (cartilage) constitute the chondrocranium and become replaced by bone to form the basicranium. Many of the fusions do not occur until childhood, a few even into late childhood, so that it is only in the human adult that they form a more or less single solid component.

6.7.3 The Dermatocranium

The dermatocranial elements also contribute to the skull. In evolutionary terms, these elements arise in the dermis (outside the deep fascia) and sink inwards to become applied externally to the cartilage bones just outlined.

Bones of the Dermatocranium

In most mammals the dermal bones come to cover most of the elements of the splanchnocranium. As most of this latter largely disappears in development, the dermal bones come to form most of the skull. This occurs early in development in humans so that teeth (that in cartilaginous fishes, for example, are embedded in the cartilages of the first branchial arch) develop in the dermal bone of the jaws.

Further, these dermal bones predominate in the formation of the skull. Thus they can be described as a series of elements forming the skeleton (i) of the lower jaw, the mandible, (ii) of the face, mainly the maxilla and zygoma, (iii) around the eyes, mainly the supporting bone of the orbit, (iv) supporting the external ear, mainly the squamous part of the temporal bone, (v) roofing the top and sides of the skull, most of the frontal, the parietal and part of the temporal and occipital bones, and (vi) most of the roof of the mouth, palate, and related components.

Muscles Associated with the Dermatocranium

The dermal bone arising in the superficial fascia close to the skin is also associated with dermal muscle arising similarly. In the trunk this dermal muscle can be extensive in mammals. In humans, however, it is at a minimum (as previously described).

The reverse is the case in the head where the dermal muscles are extensive. However, though in the trunk these muscles are not especially associated with segmentation, in the head they are closely related to the segmentation of the branchial arches.

They form a single complex group of muscles called the muscles of facial expression and are supplied by the facial nerve. They are thus developmentally distinct from the other muscles of branchial origin that are supplied by the facial nerve but that lie deep to the deep fascia (like most other striated muscles).

6.7.4 Summary

This comparative information helps in the understanding of the anatomy of the human head.

In addition however, it is necessary to understand the integral relationships between the specific nerves supplying the head structures and the arrangements of those head structures themselves. Likewise, there is a further reciprocal relationship between the specific parts of the central nervous system from which those nerves, in turn, derive. These matters are covered in Chapter 7 on the central nervous system.

CHAPTER SEVEN

Building the Human Brain

7.1 The Beginnings of the Central Nervous System

Fundamental developmental processes are involved in the elaboration of the central nervous system. To recap discussion in prior chapters, the nervous system first appears through induction of the more superficial layer of ectoderm by the underlying notochord. The dorsal ectoderm gradually forms a shallow longitudinal hollow along the midline. This deepens to become a groove, constructed of ectoderm which is, however, now specially designated as neurectoderm, and called the neural groove. The neural groove gives rise to the central nervous system. The groove descends below the surface to become a tube, the neural tube. Special cells

The Scientific Bases of Human Anatomy, First Edition. Charles Oxnard.
© 2015 John Wiley & Sons, Inc. Published 2015 by John Wiley & Sons, Inc.

from the ectoderm/neurectoderm junction on the two lateral crests of the neural groove become separated off as the neural crest. The neural tube and neural crest then undergo a series of secondary inductions that are driven by the neural tissues themselves. They come to form the brain cranially and the spinal cord caudally, together with many other special tissues (some neural and some not) from the neural crest cells. There is no simple division between brain and spinal cord. They merge into each other.

Building the human brain requires information that stems from neurobiology and neuro-embryology. These are highly complex disciplines and are usually treated quite separately from human macro-anatomy. However, there is a fundamental organization to building the human brain and spinal cord that is entirely linked with building the human head and trunk. This fundamental organization can be understood without entering into all the detailed neurobiology of what is a very complex system indeed. It is for this reason that the new findings in neurobiology are as important for understanding human anatomy as they are for understanding the nervous system itself.

Although it is more usual to describe the brain first and the spinal cord later, for the same reason, that we described the trunk first (that the trunk is simpler) and the head later, so here we describe the spinal cord first (the spinal cord is simpler) and the brain later.

7.1.1 Building Materials and Processes

Thus the central nervous system in the trunk can be described, first through the initial building components, second through the processes that modify them, and third through their effects upon the building of the extra-neural organization of the trunk.

7.1.2 Internal/External Differentiation

In a cross-section of the developing neural tube, the component cells are arranged as three groups or zones representing a first level of organization for the eventual spinal column.

The first of these is the **ventricular zone.** This comprises a layer of cells that are lining what was the dorsal surface of the neural groove and are at that point linked with the rest of the ectoderm. They are, therefore, epithelial cells and they remain so but have the capacity to give rise to new neuronal cells. With the closure of the neural groove into a tube, they become located internally, lining what becomes a fluid containing axially located channel, a narrow canal located in the center of the developing spinal cord. Thus the ventricular zone, originally dorsal in position, comes to lie internally in the center of the cord.

Outside this central layer in the spinal cord is a second component, a layer of cells called the **intermediate zone.** As the spinal cord matures, the intermediate zone

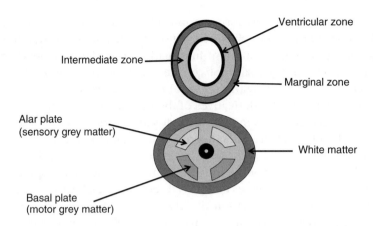

Ventricular zone

Intermediate zone

Marginal zone

Alar plate
(sensory grey matter)

White matter

Basal plate
(motor grey matter)

FIGURE 7.1

(a) Very early and (b) later differentiation of cells within cross-sections of the spinal cord.

thickens and becomes the grey matter of the spinal cord in which the cell bodies of many of the neurons are located.

On the periphery of the developing spinal cord, therefore outside even the inter-mediate zone, is the **marginal zone**. As the spinal cord matures this layer becomes the white matter. It contains the communicating dendrites and axons of the neurones migrate to the periphery of the neural tube before passing to their final destinations. This zone thus contains mainly cell processes for interconnections with other neurones, for exit and entrance to the spinal cord, but does not, in general, contain neuronal cell bodies themselves.

These beginnings are represented in Fig. 7.1(a).

7.1.3 Dorsal/Ventral Differentiation

A little later, in a cross-sectional view (Fig. 7.1(b)), the intermediate zones become organized, on each side, into what are called plates. One pair is dorsally located and named the **alar plate** (because in the brain it is a dorso-lateral 'wing'). The other is located ventrally: the **basal plate** (in the brain a ventro-medial base). These plates represent a fundamental division between dorsal sensory and ventral motor components in the spinal cord.

The organization of the alar plate involves dorsalizing influences stemming from molecular factors produced by the epidermal ectoderm adjacent to the developing neural tube. This may be in part why it is sensory in modality.

Likewise, the organization of the basal plate involves ventralizing influences based upon a series of molecular factors that are released from the adjacent ventrally located notochord. As the notochord also stimulates the pre-muscle structures that lie outside this part of the neural tube this may explain why the basal plate is primarily motor in modality.

The neurones in the alar plate form a population of cells known, in the later spinal cord, as the **dorsal horn**. Cells clustered just outside the spinal cord in the adjoining **dorsal root ganglia** transmit afferent information from the peripheral parts of the body via axons that form dorsal roots of the spinal nerves. These fibers pass into the dorsal horn and synapse with the dorsal horn cells themselves.

The dorsal root ganglia are derivatives of the neural crest and hence are also truly nervous system elements. It is in this way that sensory information enters the spinal cord through the fibers in the **dorsal roots** of the spinal nerves.

The neurones in the basal plate lie in what comes to be called the **ventral horn** of the spinal cord and the axons arising from these cells leave the spinal cord as the **ventral roots** of the spinal nerves and pass directly to the effector organs (muscles).

There is also a smaller population of cells lying between the dorsal and ventral horns called the **intermediate horn**. This contains cells that will become the spinal effector component of the autonomic nervous system. In humans, the fibers of these cells (some motor to smooth muscles in various structures, and others not strictly motor in terms of making muscles move but effector-motor in terms of making glands secrete) leave the spinal cord as a small aliquot of fibers within the mainly motor ventral roots of the spinal nerves. The sensory components that stimulate these cells stem from certain fibers entering via the dorsal roots into the dorsal horn. But the second level of neuron passes from here to the intermediate horn. These elements may be carrying obvious somatic sensory information (say about temperature or pain) or visceral sensory information (say about the state of tension of blood vessels).

The dorsal root ganglia, and indeed even the most proximal part of the spinal nerve (after fusion of dorsal and ventral roots), are covered by extensions of the meninges surrounding the spinal cord. In this sense they can be thought of as components of the central nervous system. This relationship is critical because it means that the external world (in the form of inspired air in some alveoli of the lungs at the back of the thorax) may be within a millimeter or two of the cerebrospinal fluid inside the prolongations of meninges along the proximal parts of the spinal nerves. We already saw such a close relationship in the nose where the nasal air is close to the cerebrospinal fluid within the coverings of the olfactory bulb, and in the eye where the dural sheath covering the cerebrospinal fluid connecting channels extend all the way out, along the so-called optic nerve to the retina. Even in the ear this relationship is close, the neurones of the inner ear being separated only by a very short distance (albeit by thin bony plates) from the air in the middle ear cavity.

All of this dorso-ventral organization stems from the existence of the notochord and the series of molecular factors that are released by it.

7.1.4 Integrative Components

Within both the alar and basal plates, in addition to the inputs and outflows of the sensory and motor neurons just described, are a whole series of cells and their processes that do not leave the spinal cord. These neurones (internuncial neurones, one set has already been described in relation to the autonomic outflow) are involved in communications within the spinal cord, and to and from the brain, between what would otherwise be separate cell groups. They thus have important integrating functions.

A first group of such cells and their processes are organized within individual spinal segments, into a well-defined dorsal and ventral pattern by molecular factors relating to the trunk somites on each side. They are, therefore, most evident within a single side of the spinal column. They are primarily involved in internal connections between the sensory and motor cell groups providing coordinating functions between the sensory and motor elements of the same side. For example, one system in which they are involved is the simplest of the spinal reflexes, such as the patellar tendon reflex. Another is the influence of sensory inputs into autonomic outputs.

A second group of cells and processes, especially the processes, are evident within the entire cross-section at each spinal level. These are commissural in arrangement; that is, they pass from cell groups on one side of the spinal cord to cell groups on the other, and between both sensory and motor elements. They thus also provide important coordinating functions between right and left sides, as well as between sensory and motor elements.

These two sets of fibers are associated with interactions generally within individual spinal levels or between adjacent spinal levels.

A third very large group of fibers comprises a whole series of axons that pass **longitudinally** in the developing spinal column. Some of these link different levels both near and far in the spinal cord. The great majority, however, form links between all levels of the spinal cord and the brain.

Many of those fibers carrying sensory information to the brain come from cells in the dorsal horn and pass cranially to the hindbrain in large dorsal tracts that are called the **dorsal columns**. These comprise the axons of the second level sensory neurones whose cell bodies lie in the dorsal horns.

This arrangement of these sensory longitudinal tracts maintains the dorsal part of the dorsoventral pattern that is so ubiquitous in the body generally, as well as in the nervous system. But this arrangement also maintains the cranio-caudal segmental pattern of peripheral body structures and their peripheral nerves. That is, the information from the lower limbs lies medially in these dorsal tracts, from the trunk in the intermediate region, and from the upper limb laterally.

This arrangement even maintains the superficial to deep pattern in the body generally. Those fibers from the dermatome (most superficial part of the somite, e.g. from sensory receptors in the dermis) lie most superficially, those fibers from the myotome (intermediate part of the somite, e.g. from muscle spindles) lie intermediately, and those from the sclerotome (deepest part of the somite, e.g. from pressure receptors in tendons and fascial sheets) lie deepest.

These tracts (axons) all pass cranially and terminate in the hindbrain. However, there, a second set of longitudinal fibers takes the information even more cranially to many higher centers but especially the cerebral cortex. In so doing, these fibers decussate (cross to the other side) in the brain stem. This is why, in damage to the cerebral cortex, the sensory deficit (paresthesia or anesthesia, partial or complete lack of sensation) is in the opposite side of the body.

Likewise, those long tracts transmitting motor instructions from the brain to the spinal cord lie ventrally in what are called the **ventral columns**, but more usually, the **pyramidal tracts**. The cell bodies here are located in the higher brain centers (e.g. the cerebral cortex, see later). They pass caudally to synapse with cells in the ventral horns at the appropriate levels of the spinal cord. As they descend, they, too, cross over to the opposite side (again a decussation) at the level of the hindbrain. This is why a lesion in the cerebral cortex may cause a motor deficit (paresis, partial paralysis, or complete paralysis) of the opposite side of the trunk. This arrangement of these motor longitudinal tracts maintains the ventral part of the dorsoventral pattern that is so ubiquitous in the body generally. As with the equivalent sensory elements, this arrangement also maintains the cranio-caudal segmental pattern of peripheral nerves and structures. Thus, again, the topography of the arrangement of the fibers makes sense in relation to the descending structures in the body: most cranially the upper limbs, then the trunk (thorax and abdomen), and finally the lower limbs and tail (pelvic structures and coccyx).

These levels of patterning, clearly evident in the spinal cord, become more and more complexly arranged in relation to all the new structures that are confined to the cranial end of the developing brain. Thus there are, finally, a number of other important ascending and descending axon tracts that connect the spinal cord with intermediate brain structures. The ascending tracts go to structures that develop in the hindbrain (e.g. the cerebellum, spino-cerebellar tracts), in the midbrain (e.g. the tectum, spino-mesencephalic tracts) and in the forebrain (e.g. spino-thalamic and spino-cortical). These are only three major examples of many such communications from the spinal cord to higher centers. There are equivalent descending pathways from higher centers to spinal cord from the neurones mostly located within the forebrain (e.g. cortico-spinal tracts), from the midbrain (e.g. rubro-spinal tracts), and from the hindbrain (e.g. vestibulo-spinal tracts).

All this is a level of sensory and motor coordination within the nervous system that is equally evident in the body external to the nervous system. Yet for all this complexity, almost all of these tracts are arranged internally in the nervous system in relation to the dorso-ventral, proximo-distal and ascending-descending structures of the trunk outside of the central nervous system. Thus, though knowledge of internal brain components is important in understanding brain function (neurobiology) in its own right, it also helps elucidate external anatomical structures (peripheral anatomy).

7.2 From Spinal Cord to Brain: the Initial Brain

There are elements of building the brain that more or less parallel what we have just seen in the rather simpler spinal cord. In addition, however, there are other elements that differ from the spinal cord situation.

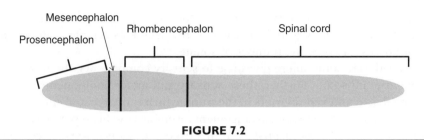

FIGURE 7.2

Major components of the very early brain.

7.2.1 The Main Parts of the Brain

Thus, even before the neural tube has formed, foreshadowing the eventual development of the enormous brain, there are three thickenings in the cranial part of the ectodermal neural plate or groove that are not present in the spinal cord more caudally. Once the neural groove has sunk below the surface as the neural tube, these enlargements become even more obvious. They are produced by molecular factors from the dorsal ectoderm and the underlying more ventral notochord and related mesoderm.

From behind forwards, these enlargements are known as the **hindbrain** (hence **rhombencephalon**), the **midbrain (mesencephalon)**, and the **forebrain (prosencephalon)** (Fig. 7.2).

These three main parts undergo further subdivision. The caudal hindbrain, the rhombencephalon, subdivides into a more cranial **metencephalon** and a more caudal **myelencephalon**. They are related to the hind part of the head and show evidences of considerable similarity to the segmental system within the spinal cord in the trunk. Thus, parallel with spinal somites in the trunk there are, eventually, somites in the head (known as occipital somites). As a result, there are non-neural elements in the head that parallel those in the trunk.

The middle **mesencephalon** remains undivided. Almost all morphological evidence of segmentation has disappeared in this region. These parts show even less relationship with the segmental structure of the trunk. For example, somites never form, but the precursors of somites in the very early trunk (see early development) remain in this part of the head as the doubled segmental elements called somitomeres. Molecular biology shows that, at the molecular level, somitomeric segmentation is in fact present and also affects non-neural elements of head structure.

The prosencephalon, the most cranial part, divides into a **telencephalon (endbrain)** with prominent outpocketings that ultimately form the cerebral hemispheres, and a more caudal and deeper **diencephalon** (which already shows the optical vesicles lying laterally). These parts show even less relationship with the segmental structure of the trunk.

The difference seems to be based upon retention in the head of the system of somitomeres in the trunk. In the trunk the cranio-caudal pairs of somitomeres fuse very early into somites. This influences the positions of the spinal nerve roots so that there is one pair (sensory and motor) for each spinal segment. In the head, but somewhat differently at different levels, the segments remain mainly separate as somitomeres (at twice the number of somites).

In the head, one somitomere gives rise to sensory components, the next to motor components, and so on. Further, these sensory and motor components do not join (as do the dorsal and ventral roots in the trunk which form a mixed spinal nerve). It is of interest that this segmental arrangement resembles the main pattern of segmentation in creatures like the hagfishes. In hagfishes, the dorsal sensory components are out of register and separate from the ventral (motor) elements. This ancient comparative story may actually be behind one of the otherwise puzzling aspects of difference between the trunk and the head (see later).

7.2.2 The Folding of the Brain

These five subdivisions of the very young brain represent a fundamental organization that persists throughout life. They increase in size very rapidly at these early stages of the embryo, and as a result become folded upon one another. A first fold, convex dorsally, and called the **cephalic flexure**, occurs in the midbrain (mesencephalon) region. A second fold (also convex dorsally and called the **cervical flexure**) occurs at the junction of the hindbrain with the developing spinal cord. A third, reverse, fold (convex ventrally and called the **pontine flexure**) occurs between the two (Fig. 7.3).

The result of this folding process in the brain is that, contrary to the situation in the spinal cord where segmental structures follow in a relatively straight linear sequence from cranial to caudal, the structures in the developing brain, though sequential, are arranged in a curvilinear fashion, along the length of a '3' lying upon its back as it were. This influences our view of the anatomy. In the spinal cord, the anatomist (or pathologist) can easily make transverse sections that are all at right angles to the linear arrangement and therefore display the structure of a single segment. In the brain, because of the folding, the transverse sections of the anatomist (or pathologist) are not at right angles to the true longitudinal axis. As a result various structures weave in and out of the sections in a manner that hides their fundamental serial organization. This makes the pattern of brain arrangements difficult to understand.

Yet the brain structures are truly serially arranged as in the spinal cord. Thus, the clearest way to understand their pattern, their positions, and their functional relationships, is to conceptualize an arrangement where the folding is (diagrammatically) unfolded (Fig. 7.4). We must also make allowance for this when we examine standard sections.

7.2.3 The Earliest Building Components, Neuromeres and Their Segmentation

As with all other parts of the body, but rather more complexly, we can now discern the primary building components from which the brain derives. However, the

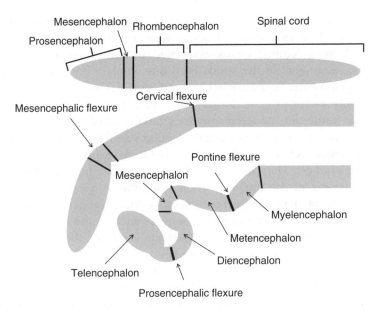

FIGURE 7.3

Changes in position of major brain components as various brain folds occur during development.

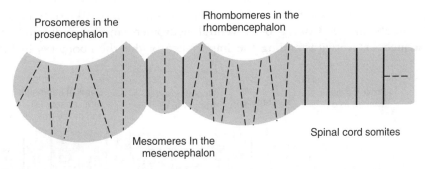

FIGURE 7.4

Internal segments (neuromeres) of the main components of the brain (in which the folding has been undone for clarity) compared with the spinal cord.

changes are so rapid and complex that it is clearer to examine the building components alongside the building processes.

Thus the brain develops in relation to a set of longitudinal segments called 'neuromeres' (Fig. 7.4). These seem to be the neural equivalents of the extra-neural somitomeres evident in the head, outside the brain. This contrasts with the trunk; the extra-neural somitomeres in the trunk are transient and are almost immediately converted into somites. The equivalents of neural 'somitomeres' never appear in the spinal cord. The first indications of segmentation in the spinal cord are the alar (sensory) and basal (motor) plates associated with the appropriate extra-neural trunk somites. Segments in the spinal cord thus contain both dorsal sensory and ventral motor elements. Segments in the brain remain separate; each alternate neuromere has only its appropriate dorsal (sensory) or ventral (motor) components.

Thus, in the hindbrain, underlain by the cranial end of the notochord, the neuromeres are strongly retained as a highly regular segmental organization. They are here called **rhombomeres** (r) and are numbered from cranial to caudal (r1 to r8, and possibly there is an r9). They are arranged as a sequence, the first, r1, being dorsally located, the second, r2, ventral, and so on along the series (Fig. 7.5). (In this dorsal/ventral arrangement they reflect the dorsal/ventral arrangement seen throughout the nerve cord in much more primitive creatures such as the hagfishes). These not only determine the organization of the hindbrain itself, but they also presage much of the overall organization of the face and neck.

The correspondence between the rhombomeres of the developing hindbrain and the somitomeres of the peripheral structures of the head: branchial arches, cranial glands, and so on, is remarkable. For instance, the afferent (special visceral sensory) fibers in cranial nerves **V**, **VII** and **IX** develop in relation to rhombomeres 2, 4, 6, and 8, and innervate structures from the 2nd, 4th, and later somitomeres. These rhombomeres are more dorsal in position.

Equally, cranial nerves **IV**, **VI** and **XII** develop in relation to ventral rhombomeres 1, 3, and 7 and innervate somatic muscular structures (muscles of the eyeball and tongue) via somato-motor fibers. (The related cranial nerve **III** arises in relation to the midbrain but from a prior ventral neuromere in the midbrain that fits this sequence). Notwithstanding the importance of the rhombomeres in helping

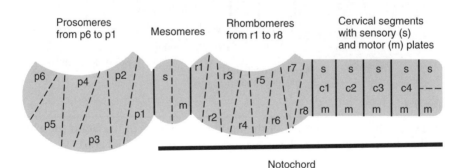

FIGURE 7.5

Internal segments numbered as prosomeres (p1, p2, etc., in the forebrain), mesomeres (not numbered in the midbrain) and rhombomeres (r1, r2, etc., in the hindbrain) showing their dorsal and ventral alternation. These can be compared with spinal cord segments which contain dorsal, s, sensory, and ventral, m, motor components within each single segment, relating to each individual somite in the trunk.

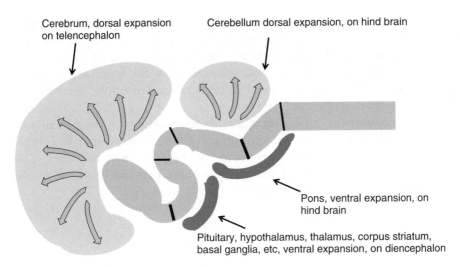

Cerebrum, dorsal expansion on telencephalon

Cerebellum dorsal expansion, on hind brain

Pons, ventral expansion, on hind brain

Pituitary, hypothalamus, thalamus, corpus striatum, basal ganglia, etc, ventral expansion, on diencephalon

FIGURE 7.6

A return to the folded brain showing the dorsal and ventral expansions that occur in each brain region.

determine the course of development, they are in fact fairly transient and soon disappear, leaving, in their place, the occipital somites.

The various neuromeres also contribute to later dorsal expansions of new brain parts.

In the hindbrain, the rhombomeres contribute to a dorsal expansion that forms the cerebellum, and a ventral expansion that contains a number of hindbrain nuclei and the pons (Fig. 7.6).

In the midbrain the segmental structure is much less evident (Figs 7.5 and 7.6). Such neuromeres as there are disappear almost as soon as they are formed, leaving only one dorsal and one ventral mesomere. The ventral one is that associated with the **III** cranial nerve in the ventral sequence with **IV**, **VI** and **XII** already discussed. Yet this does not mean that the midbrain is not segmented. The segmental patterning is still present but this has been shown to be only at the molecular level.

In the forebrain, neuromeric segmentation is again evident as a set of six neuromeres but in this region they are called prosomeres (Figs. 7.5 and 7.6). These extend from the tip of the forebrain cranially (or rostrally) to the midbrain/forebrain boundary more caudally. Unfortunately, and as a result, confusingly, they are numbered in a different way, in reverse order from p6 at the cranial end to p1 at the caudal end. Again these alternate, dorsal and ventral.

The more dorsal prosomeres contribute to the dorsal expansions of the fore-brain (e.g. the future cerebral cortex, corpus callosum, pineal gland, etc., Fig. 7.6). The more ventral contribute to deep ventral structures in the forebrain (e.g. the neural part of the pituitary gland, the related hypothalamus, the corpus striatum, and the thalamus, this last a major relay station between the cerebral cortex and the body, etc., Fig. 7.6).

A further series of structures become apparent. These are the **olfactory**, **optic** and **stato-acoustic** vesicles that, together with related epidermis, come to form the special sense organs of **smell**, **vision** and **hearing/balance,** respectively. These vesicles arise, one each, in the three main brain enlargements, the forebrain, midbrain and hindbrain. Their nerves are numbered, in the sequence of cranial nerves, as nerves **I**, **II** and **VIII** (as described in Chapter 6).

All these arrangements show, yet again, the dorso-ventral arrangement so ubiquitous in development. However, the full story behind these transient, but organizationally important, components is not yet known. That there are, however, correspondences between brain and spinal column, between head and trunk, and across many species, is not in doubt.

7.3 The Ultimate Brain

In the spinal cord were described a series of further developments that relate to neural coordination within the cord. The first of these involved the development of internuncial neurones, passing between those neurones and their processes that have external links to and from the structures outside the spinal cord. The second involved the development of commissural fibers that cross the midline to coordinate the activities of the two sides of the spinal cord. The third involved the development of a series of longitudinal fibers, passing up and down the cord, linking various levels of the spinal cord, but especially linking the entire spinal cord with the different longitudinal levels of the brain. These patterns are also evident in the brain although rather more complexly.

Hindbrain in General

The intrasegmental separation into alar and basal plates is every bit as evident in the brain as in the spinal cord. However, what was the small central channel in the spinal cord is, in the hindbrain, widened out into a broad channel called the fourth ventricle. It also is more dorsally placed rather than being buried centrally as in the spinal cord. As a result, instead of the alar and basal plates being simply dorsal and ventral in position around a narrow central channel, they are dorso-lateral and ventro-medial, respectively, underneath the broad dorsal ventricular estuary Fig. 7.7, compare Fig. 7.1(a,b)).

As a result, in the hindbrain, the various components that existed in the spinal cord as dorsal and ventral in position on each side are here changed in position, being dorso-lateral and ventro-medial, respectively. Further, in relation to the additional functional components of the head, the dorso-lateral and ventro-medial components show additional subdivisions.

Thus, in the alar plate, (dorsal in the spinal cord but dorso-lateral here) are cell nuclei. One, most dorsolateral, is the nucleus of neurones for the special senses (in Fig. 7.8, at this level for hearing and balance). These are called special somatic

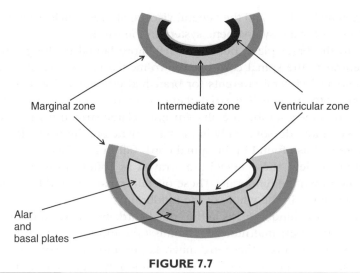

FIGURE 7.7

Initial zones on a cross-section of the very early brain.

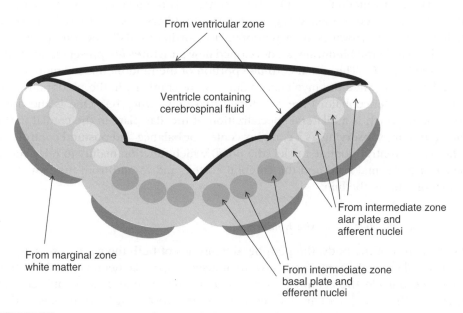

FIGURE 7.8

Later components developing from the zones of Fig. 7.7. (Compare with spinal cord Fig. 7.1(a,b).)

afferents. Arrayed more medially from this nucleus (also Fig. 7.8) are the general somatic afferent nuclei (dermal sensation, equivalent to that in the trunk), the special visceral afferent nuclei (branchial afferents such as new sensory inputs, e.g. taste from the branchial arches) and the general visceral afferent nuclei (equivalent to the sensory afferents for the sympathetic system in the trunk).

Likewise, in the basal plate, but somewhat more lateral to the general somatic efferent (ventral in the spinal cord but ventro-medial here) develops a nucleus of cells called special visceral efferents (or branchial efferents). These are associated with the new motor outputs (e.g. striated muscle of the branchial arches) around the cranial portion of the gut tube, the pharynx and related structures, also not found in the trunk. Even more dorsolaterally in the basal plate are the nuclei for the general visceral efferents (also found in the spinal cord) (all in Fig. 7.8).

In longitudinal view, these nuclei are arrayed as columns on each side as one passes up and down the hindbrain. These overall transverse and longitudinal organizations are evident in the whole of the hindbrain.

Further, within the hindbrain, as in the spinal cord, are the various intersegmental coordinating structures: internuncial, commissural and longitudinal fiber bundles. In the hindbrain, however, these are much larger and more complex than in the spinal cord being evident as cross-sectional and longitudinal tracts. Those involved in sensory functions lie, as would be expected, in the alar dorsolateral region, and are carrying information into higher brain centers. Those involved in motor activities lie in the basal ventro-medial region and are the fibers passing caudally from the higher centers downwards to the hindbrain and further caudally into the spinal cord.

Finally, within the hindbrain are developed new structures not present in the spinal cord (see Fig. 7.6). The spinal-cord-like portion of the hindbrain is the more caudal medulla (often called the myelencephalon). More cranially, in the hindbrain (often called the metencephalon) are developed two new structures. One is the dorsal cerebellum which is a bilateral specialization of the alar plate in the cranial part of the hindbrain, subserving as a complex center for balance and postural control. The other is the ventral pontine nuclei (the pons) which function mainly to relay signals between the spinal cord and the cerebral cortex, and to and from the new dorsal structure, the cerebellum.

Afferent Ganglia Related to the Hindbrain

As in the rest of the body, the peripheral neurones of both the general somatic and the general visceral afferent pathways are housed in ganglia, neural crest derivatives, which lie outside the central nervous system. The general somatic afferent cells related to the spinal cord are located in the dorsal root ganglia. In the head, these are mainly in the very large ganglion of the **Vth** cranial nerve. This is the major (but not the only) part of somatic sensory perception in the head. Much smaller numbers of general somatic afferents are also found in smaller ganglia associated with the **VIIth**, **IXth**, and **Xth** cranial nerves. They are organized linearly but not as a clear chain of ganglia as in the trunk.

7.3.1 Hindbrain Pattern Modified in the Midbrain

In the midbrain, those elements of spinal cord structure that are evident in the hindbrain are even less obvious. The midbrain is primarily a relay center connecting the forebrain, the hindbrain and the spinal cord. Its biggest components, therefore, are the aforementioned massive longitudinal connecting tracts. It does also contain, however, several important cranial nerve nuclei. Two of these are, as described earlier, cranial nerves **III** and **IV** (and a related cranial nerve **VI** lying slightly more caudally). They subserve mostly the external eye muscles. They lie in the ventro-medial position of a typical somatic motor nerve nucleus.

A third nucleus, that appears to be in the mesencephalon, is a cranial portion of the already mentioned sensory portion of the **Vth** nerve nucleus. This lies in the dorso-lateral position of a typical general somatic sensory nerve nucleus. It is probably, however, really part of a nucleus that belongs in the hindbrain but that has been secondarily displaced across the boundary into the midbrain.

The position of these nerve nuclei emphasize that the fundamental relationships between the alar and basal plates are preserved. They are largely based upon the same molecular signals (e.g. sonic hedgehog) as in the spinal cord and hindbrain. Similar working arrangements are also present with the dorsal parts of the midbrain being associated with generally sensory functions and the ventral parts with generally motor functions. But the structural arrangements within the alar and basal plates in the midbrain are less evident. This may be mainly because of the primary relay functions of the midbrain as well as new special functions associated with the special senses of the head.

Thus the alar plate in the midbrain forms the sensory components as before. But it largely subserves the functions of vision and hearing, the two suprasegmental sensory organs that arise very early in the head in relation to previously described optic and otic (stato-acoustic) vesicles (and later capsules). Thus two paired bulges in the alar plate (colliculi) are functionally involved in these special senses; the cranial-most pair being involved in vision, and the caudal-most in hearing.

The equivalent special function of smell (olfaction) is related to a development that is yet more cranial than the midbrain in developing olfactory brain lobes related to the previously described olfactory vesicles (and later, capsules). These structures are forebrain developments, the olfactory lobes and tracts. The other special sense is that of pheromone detection, like smell but highly specialized, in relation to the terminal nerve and the vomero-nasal organ and it is also primarily a forebrain derivative.

Finally, within the midbrain are the various commissural and longitudinal features already described in their passage to and from the spinal cord and the various higher centers. Likewise, the narrow central spinal canal that opens up into the wide dorsal canal of the hindbrain narrows once again in the midbrain into the narrow cerebral aqueduct connecting with the much larger ventricles of the forebrain.

7.3.2 The Ultimate Complexity of the Forebrain

The structure of the forebrain is even more complex than any of the more caudal regions (see Figure 7.6). It was previously believed to be composed only of an expanded alar plate. However, there is considerable information that suggests that, just as in the rest of the system, there are also basal plate components. Like the more caudal parts the organizational components are neuromeres, here called prosomeres as already explained, but numbered in reverse order from p1 to p6.

Alar plate derivatives (i.e. dorsal in position) of the diencephalon proper are the epithalamus and the epiphysis (pineal body), a phylogenetically primitive gland that often serves as a light receptor (sensory) In humans it is probably mainly associated with daily light–dark cycles of activity and the secretion of melatonin.

Basal plate derivatives of the diencephalon become transformed into the various parts of the thalamus, a major relay station for transmitting signals between the cerebral cortex and the more caudal parts of the nervous system. They also include the hypothalamus and the neurohypophysis for integrating autonomic nervous functions and endocrine release from the pituitary.

The telencephalon, the second part of the forebrain and a dorsal expansion from the diencephalon, comprises, eventually, the cerebral cortex and its related but more internal structures, the palaeocortex and corpus striatum. In humans, this part of the brain comes to be enormously expanded so that it ultimately overgrows other parts of the forebrain and comes to have the grey/white matter reversed, compared with the rest of the nervous system. That is: the grey matter is externally placed and the white matter internally. In this it is like the cerebellum that also overgrows the other parts of the hindbrain and is also organized with its grey matter (cells) externally and white matter (fibers) internally. Such growth and structural reversal provides for a much greater amount of brain cellular matter in both of these structures (again, see Fig. 7.6).

These particular brain components can also be named through their evolutionary development. The oldest (in evolutionary terms) is called the rhinencephalon (also called the palaeocortex). As the name implies, it is heavily involved in olfaction (but also, in humans, many other functions). The cerebral cortices (also called the neocortex), at first smaller than the palaeocortex, increase even more markedly during development so that they become most of the mass of the brain.

Longitudinal Arrangements in the Forebrain, Neural Organization

Thus, the fundamental relationships between the alar and basal plates are only somewhat preserved in the forebrain. They are largely based upon the same molecular signals (sonic hedgehog) as in the spinal cord, hind- and midbrains. But the structural arrangements within the alar and basal plates in the forebrain are distorted because of the enormous developments of the alar derivatives, the cerebral cortices (p4-6) and the fact that this results in p4-6 folding over, therefore almost burying p1-3.

Further, it is entirely possible that the numbering of the prosomeres provided by developmental neurobiologists hides the true relationships. It seems not impossible that the prosomeres, p1-6, are actually alternating prosomeres, one basal, one alar, along the length of the developing forebrain (as in the other parts of the developing brain).

For logical consistency they should also be numbered in reverse order (within the forebrain) from cranial-most (p1) to caudal-most (p6) as in the whole of the rest of the nervous system. This would make the homologies, and therefore the relationships of both internal neural arrangements and external head structures much clearer. However, until further studies of developmental and neurobiologies clear these matters up, we must, of course, stick with the current terminology. This is yet another example of how anatomy provides hypotheses for development and neurobiology.

Longitudinal Arrangements in the Forebrain, Central Canal

Alongside the expansion of the forebrain, especially the cerebral cortices, is an equivalent expansion of the small midline central canal of the spinal neural tube. In the brain there are large lateral expansions of the central canal and these are connected to the central channels by smaller channels. These are the complexly-shaped lateral ventricles filled with cerebrospinal fluid. This is produced by specialized areas called the choroidal plexuses, highly vascular regions that project into the ventricles and secrete the cerebrospinal fluid.

Longitudinal Arrangements in the Forebrain, Special Senses

As already noted, the forebrain is also involved in the development of certain special sense organs that are unique to the chordate head. These might at first be thought to be matters for the building of the head; they seem to be external to the brain. They certainly do influence in great measure the anatomy of the head external to the brain. However, though the major sense organs arise in large measure from the ectoderm, they arise from regions of thickened ectoderm that appear lateral to the neural plate in the very early embryo. There is, thus, an interaction in the development of the peripheral head structures and the central nervous system.

7.3.3 Overall Brief Summary: Relationship between Brain and Head Development

The earliest obvious segmentation in the trunk appears in the non-neural structures. It is only later that segmentation in the spinal cord becomes obvious.

In contrast, in the head, segmentation appears very early in the developing brain. It is associated with the developing segmentation in the non-neural parts of the head (see Figs. 6.16 and 6.17 in Chapter 6).

7.4 The Size and Complexity of the Brain

Perhaps the most striking immediate element of the anatomy of the brain in humans, especially in comparison with many other animals, is its enormous size, and its effect upon the size and structure of the head. Many studies have been carried out to try to understand this phenomenon. For example, the sizes of brains of different groups of vertebrates show quite different scaling factors.

7.4.1 Overall Brain size

One can scarcely mix up brains of current day fish, amphibia, reptiles and mammals (Figs. 7.9 and 7.10). They all evince quite separate patterns.

However, when we put brain size into the time dimension, it is evident that the brain sizes of ancient mammals (calculated from the brain volumes of their skulls) are neatly intermediate between those of modern reptiles and modern mammals (Fig. 7.11).

In examining brains in primates alone, those creatures evolutionarily closest to humans, we discover (apparently) that there is very little difference among them.

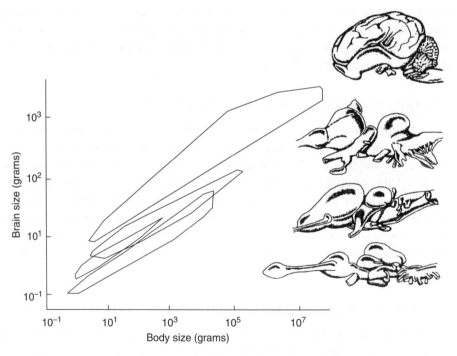

FIGURE 7.9

Brain size body size ratios in various vertebrates identified by cartoons of the different brain forms (fish, amphibian, reptile, bird, mammal).

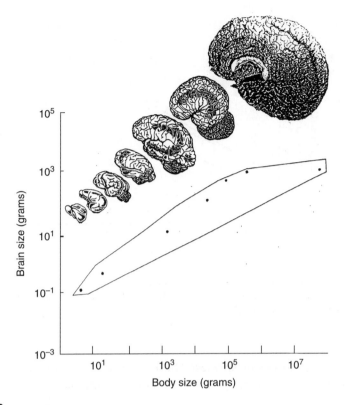

FIGURE 7.10

Brain size body size ratios in various mammals identified by cartoons of the different brain forms (armadillo, hare, dog, chimpanzee, human, dolphin, blue whale).

Ninety-six percent of what difference there is seems to be explicable on the basis of body size alone (Fig. 7.12 (a, b)). The relationship seems linear!

When, further, we try to include the time dimension by adding fossils, we seem to see a continuation of that concept, the prehuman fossils seem to lead up from australopithecines (close to chimpanzees) to modern humans (Fig. 7.13). However, looking more carefully at how such data are analyzed suggests that all is not quite what it seems; there is more than one way to envisage the relationships (Fig. 7.14). Applying these ideas to the real data from the different fossils suggests that, in relation to human chimpanzee differences, there has not been a simple linear relationship with time. Rather, there may actually have been several different groups that do not support the concept of a simple linear increase in brain size (Fig. 7.15).

As a result, we need to ask the question: can it really be that, of all the details of gross brain anatomy, brain size is the main difference between humans and other primates, especially between humans and our closest living relatives, chimpanzees? And are we certain that brain evolution has proceeded in a progressively increasing

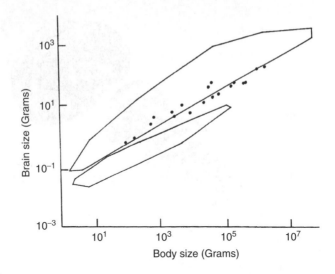

FIGURE 7.11

Brain body size ratios in reptiles and mammals, with the ancient mammals (separate points) lying neatly between them.

manner from older to newer species? These questions lead us to the examination of how brain part sizes change: that is, the effect of internal differences within the brain.

7.4.2 Brain Part Sizes and Interrelationships

At first sight, studies of brain part sizes seem to give the same answer as whole brain sizes; and a very ordinary result it is. It is a result that says: yes, the primary relationship (96% or more of the information) of brain part size is with overall brain size and with body size (Fig. 7.12(a, b). Are, then, the great cognitive differences between chimpanzees and humans very little to do with the anatomical structure? Are human brains anatomically really just ape-like brains writ larger? The common sense of these studies of brain size seems to say yes. But then commonsense is often wrong.

It is possible that this is not the whole story.

It turns out that we can test that idea from the anatomy of brain parts (Fig. 7.16). If the examination of the sizes of brain parts is carried out with a better mathematical tool that is able to ask questions about relative differences within, as compared to between, different species (a method known as canonical variates analysis) then we get new answers (de Winter and Oxnard, 2001). On such a basis, humans are a remarkable 30 standard deviation units different from any primate, including chimpanzees. And to place this in context, chimpanzees, and all the other apes, lie within only 4 standard deviation units of each other.

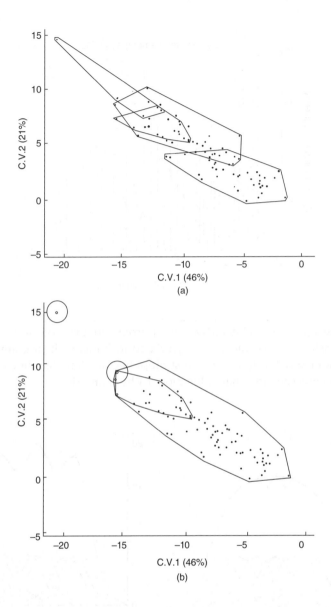

FIGURE 7.12

(a) Brain part relationships throughout the primates. The relationship is a straight line, almost totally related to body size, from strepsirrhines (right hand polygon), through new world monkeys (next polygon), old world moneys (next polygon), to apes and humans (left most polygon). Humans are the small isolated point at the extreme top left. (b) More detail about the unique position of humans. The large polygon is all non-human primates. The next smaller polygon is all Old World primates. The two circles are chimpanzees and humans and these are 8 standard deviation units apart – almost as much as covers all apes and Old World monkeys.

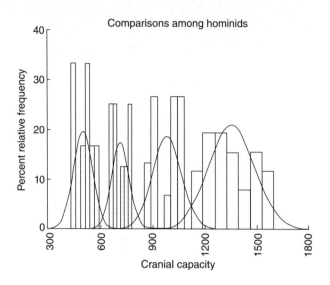

FIGURE 7.13

Brain size distributions (estimated) on an approximate time axis in modern humans and those fossils believed most closely related to humans: histograms and normal curves of endocranial volumes for australopithecines left, habilines next, erectus next, and humans (neanderthals plus modern humans) right.

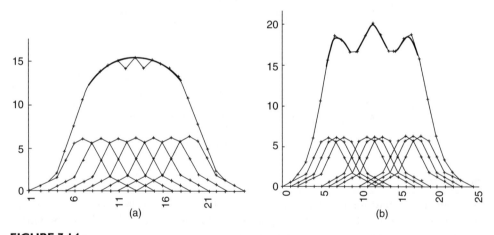

FIGURE 7.14

Addition of distribution representing continuous change over time gives an overall distribution like that in (a). Addition of distribution representing discontinuous change over time gives an overall distribution like that in (b). What is the overall distribution for the real data? See Fig. 7.15.

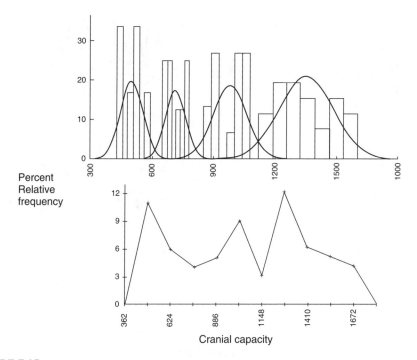

FIGURE 7.15

Addition of the separate distributions for each fossil species provides an overall distribution for the fossils with a peak for each fossil form. This resembles Fig. 7.14(a) implying that the fossils do not have a simple linear relationship. Each fossil is separate, within the limits of the rather small sizes of the samples and the approximate dating of the fossils.

What is the cause of this enormous human/non-human, human/chimpanzee, difference?

It turns out that 8 of the 30 standard deviation units separating humans from chimpanzees involve differences in the relative sizes of brain structures that are an extension of the trend common to all other primates. Eight standard deviation units is quite a lot but it is just a prolongation of a difference already evident across the non-human primates and seems indeed to be related to size (Fig. 7.12(a,b)). This trend primarily involves, as we would have guessed from the prior discussion, increases of certain very large alar brain elements (neocortex, basal ganglia, diencephalon and cerebellum) all relative to the hindbrain (Figs. 7.17 and 7.18).

But the remaining 22 standard deviation units of difference between humans and chimpanzees are produced by statistical combinations of brain parts that apply only to humans and are not found in any other living primate (Fig. 7.19). These involve interactions (including triple and quadruple loops) among structures that all lie within the brain. They include components of midbrain, diencephalon, palaeocortex

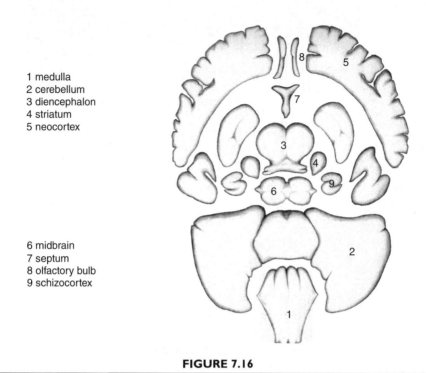

1 medulla
2 cerebellum
3 diencephalon
4 striatum
5 neocortex

6 midbrain
7 septum
8 olfactory bulb
9 schizocortex

FIGURE 7.16

A diagram of the parts of the brain.

and neocortex among others (Fig. 7.19). These relationships are not evident in any non-human primates.

This extra separation is even more significant when we realise that the particular brain part variables available in these analyses do not include any of the brain part differences *within* the neocortex; we have no data for within-neocortex differences. If we only had such data, of the relative sizes (and perhaps therefore importance) of the frontal and temporal cortices, for example, how much more different again might the human brain be from that of other even closely related primates? However, other neurobiological studies of cortical function and developmental brain chemistry are providing much information about this in humans though we do not have many comparable data from non-human primates for comparison.

Such internal cortical interactions should reflect the functions that are so much better developed in humans than in non-humans: increased levels of communication, increased information processing, increased planning and control strategies, increased abstract thought. This is only a beginning 'wish list'. If only we could partition anatomical information within the cortex, how much more different again from chimpanzees, might we be?

Can we obtain any information about this problem from the human and pre-human fossils? Of course, looking at brains in fossils is problematical. But

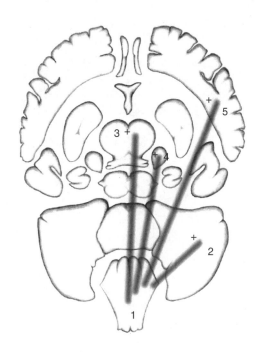

1 medulla
2 cerebellum
3 diencephalon
4 striatum
5 neocortex

FIGURE 7.17

A diagram of the connections between the parts of the brain that produce the 8 standard deviation units of difference between chimpanzees and humans shown in Fig. 7.12. These parts all relate to 'medulla/higher center' relationships, are the same for all primates, and involve only pairs of brain parts.

the tight developmental relationship between, on the one hand, superficial brain structure and size and, on the other, internal skull structure can be used to test some hypotheses about human brain evolution. Thus, studies of endocasts in a range of human fossils indicate that modern humans have, for instance, larger temporal lobes, wider frontal lobes, and enlarged olfactory bulbs (this last is apparent in Fig. 7.19) These brain features seem to be unique to modern humans, even when compared with earlier human species. Might such features relate to learning and social abilities, to memory and language processing, to the attribution of mental states to others, and, possibly, to changes in olfaction-related mating-behavior with immunological improvement of offspring?

7.4.3 Additional Aspects of Brain Size

It is also possible that these findings on brain part volumes have another, quite different, functional implication. The data analyzed here all relate to the volumes of brain parts with the idea that it is the bulk of neuronal cells and cell processes that contribute most to brain volume. We can, however, also ask the question: are

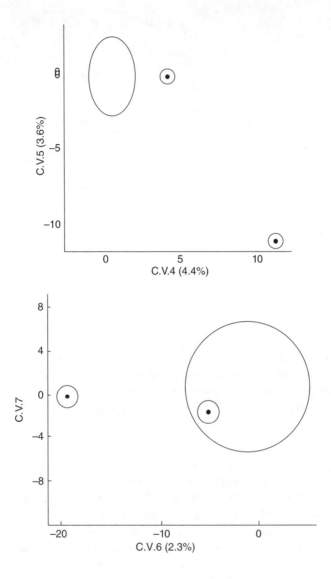

FIGURE 7.18

The relationships among non-human primates in several additional canonical variates (two bivariate plots are required to demonstrate these four statistically independent relationships). These species are contained in a very tight space (the large circles). The positions of humans are given by the small distant circles. In total, humans are a 22 standard deviation unit outlier. The position of the chimpanzee is given, as a reference, by the small circle inside or close to the large one. In other words there are additional canonical axes that do not separate any of the species save humans.

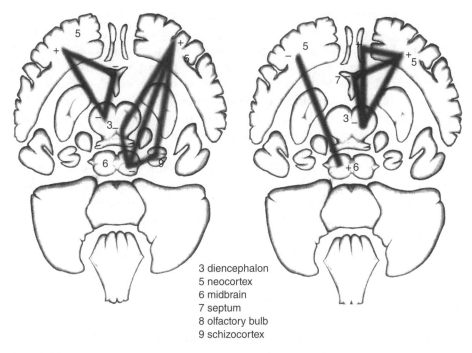

3 diencephalon
5 neocortex
6 midbrain
7 septum
8 olfactory bulb
9 schizocortex

FIGURE 7.19

A diagram of the parts of the brain that produce the 22 standard deviation units of separation in Figure 7.18 and that do not differentiate any non-human primates. These relations are 'within-brain relationships', are found only in humans, and some involve relationships between more than two brain parts.

there any other components of the brain that contribute largely to brain and brain part volume? The answer is, of course, yes. These volumes are not only due to neurons and their processes, but also the various supporting cells (glial cells) and their processes. Glial volume occupies much more of the brain in humans, as much as 40% more, than in chimpanzees.

How important could this be? The older view of glial cells was that they were primarily nerve-cell-supportive. This is almost certainly true. They are involved in such activities as bringing nutrients from the blood to the nerve cells, maintaining appropriate ion balances in the nerve-cell microenvironment, protecting nerve cells immunologically against pathogens, and specifically, for special glial cells (oligo-dendroglia centrally and Schwann cells peripherally) forming and maintaining the insulating myelin sheaths of axons of many nerve fibers.

However, additional new ideas about glial function are now abroad. Thus, glial cells appear to engage with nerve cells in a continuing two-way dialogue from the embryo to old age. This dialogue influences the formation of new synapses at well-used nerve cell contacts and the elimination of old synapses at unused contacts.

It thus helps determine which nerve–nerve links get stronger and which weaker. The glial cells also seem to communicate among themselves (through chemical rather than electrical mechanisms) as a separate but parallel communication network to the electrical nerve cell network. They talk with different local pools of nerve cell contacts over considerable distances. As a result, through them, the synaptic activity of one nerve cell pool can influence development of synapses in another without there being a direct electrical link between them. In effect, the glial cells act as a new kind of chemical-communicating mechanism both among themselves and between nerve cell populations.

It has long been known that there are relatively more glial cells per volume of brain tissue in larger as compared with smaller brains. When first discovered, this fact was laid at the feet of brain geometry. The density of nerve cell bodies decreases as a function of brain size. Some of this extra space is occupied by more and longer nerve cell processes. Because cell size increases only slightly with body size, there must, therefore, be more glial cells per nerve cell in larger brains. At a time when the only functions of glial cells were supportive this did not seem to especially affect nerve cell function. However, if glial cells also function in changing nerve cell connectivity, then this may be a different matter.

If this effect on formation of new synapses continues throughout life, it could have very important ramifications in pubertal, adult and even, and especially, aging brains. However, even if this new view of glial function should turn out to be incorrect, the metabolic implications for brain function must still be extremely important. A new brain might be expected to require a new metabolic support.

These glial possibilities are rendered even more interesting by new studies of the genes and molecular systems involved in brain size. One specific gene form (SIGLEC 11, a particular sialic acid binding receptor) is known to be unique to human brains (unique, that is, in being absent in chimpanzee, bonobo and orangutan brains). This factor is especially evident in human microglial cells. Could such a factor be involved in the production of the relative and absolute increase of glial cells in humans? Would this affect brain connectivity and function through adding and subtracting nerve cell links over physiological time? Would this, in turn, produce the flexibilities and changes that are now known, in humans, to extend into maturity, and even into old age? This could even help explain how surprisingly well some older brains can cope with some of the neurological insults acting on the brain through damage and time.

Even more interesting, could the relative and absolute increase in glial cells, through adding and subtracting nerve cell linkages over evolutionary time, produce the flexibilities and adaptations in which modern humans are clearly different from other primates? Could this account for part of our finding of the unique separation of modern human brains from other primate brains? Is it possible that all this glial involvement in modifying, supporting and extending nerve cell connectivity is part of the reason why such crude data as relative brain part volumes are able to provide the rich lifestyle associations demonstrated here?

7.4.4 Are There Yet Other Brain Components Contributing to Brain Size?

Let us return again to the question prefacing the discussion of glial cells. Are there any other components of the brain that contribute largely to brain and brain part volume?

Congenital malformations of the brain may provide useful clues here. Mutated gene systems that cause brain reductions imply the possibility of other mutations that produce brain enlargements. These might be part of the business of growing a brain during development. Could such effects spill over into evolution? Certainly, within the last few years many genes have been mapped that seem to be involved in building a bigger brain.

At least eight genes help to create the very large human frontal neocortex. Deleterious mutations in these produce smaller and thinner neocortices than normal. Children with the condition walk slowly, and are inarticulate and clumsy. Opposite mutations in such systems might cause particular brain regions, like the frontal lobes, to enlarge.

Further, some families have microcephaly genes, and their offspring are born with tiny brain cases and brains, and generally suffer from mental retardation. Their brains look fairly normal but are closer in size to those of chimpanzees and gorillas. Could opposite changes in microcephaly gene systems have been involved in the evolution of the large brain of humans?

Much more specifically, a particular gene (ASPM) is active in the proliferative regions where nerve cells which eventually migrate to other areas of the brain are born. One region or domain of this gene is repeated more often as brains grow larger. Thus, a simple worm has two copies of the region, a fruit fly 24, a mouse 61, and a human 74. One might think that the changes in the ASPM gene could be partly responsible for the very large human cortex. Other work, however, implies that, among primates, this idea may be spurious; macaques, chimpanzees and gorillas, all with smaller brains, have the same number of repeated ASPM genes as humans.

More recent work on ASPM shows, however, that variants of it are indeed specific regulators of brain size, and that their evolution in the lineage leading to *Homo sapiens* has been driven by very strong positive selection. This applies particularly to one genetic variant (haplogroup D haplotype 63) that may have arisen as recently as only 6000 years ago. Another variant, microcephalin (MCPH1) that also regulates brain size, seems to have evolved under very positive selection, and arose in the human lineage about 37 000 years ago. Undoubtedly, there have been others over the period since humans separated from chimpanzees. Perhaps (like the myosin heavy chain in terms of muscle reductions, the MYH 16 of a prior chapter), uniquely human molecules for brain increases started as long ago as 2 million years or even more, though it would seem obvious that exponentially faster and faster changes have occurred in the most recent times.

It seems as though we are close to understanding these various processes. It may soon be possible to tell which genes, which molecules, which mechanisms, and which interactions, are crucial to the human brain. In particular, we are close to understanding the development of mammalian brains as distinct from vertebrate brains generally, of primate brains as distinct from mammalian brains generally, and, and this will be critical, of modern human brains as distinct from other primate brains, even from those of chimpanzees, our closest living relatives in the supposed human evolutionary story. We seem so close to the chimpanzee in body, so different in brain.

7.4.5 Evolution of the Human Brain: New Mechanisms?

All the above ideas imply that the matter of the size of brains and brain parts is not just a crude measure. Perhaps increase in size *per se* is a merely a primer that permits various kinds of complexity to occur. Perhaps size, and especially relative size, is more important than we think. Brain size, brain part sizes, and especially relative brain part sizes, after years in the doldrums, may once again come to take a more prominent part in the understanding of brain development and evolution. Certainly, the new surprise in these results is the evidence that they offer of major differences between humans and chimpanzees.

This very large difference leads to new questions. Could it be that, in humans, changes in brain organization are partly due to new evolutionary processes? Do current views of evolutionary mechanisms allow such a question? Are there new processes that could permit, indeed produce, very fast brain change, great elaboration of function and behavior, and totally new functional (lifestyle) possibilities? Such changes in brain organization could stem, at least in part, from elaborations of the brain attendant upon mechanisms for forming new connections that are not available even to our closest living relatives, the great apes.

It is well known, that increased levels of brain inputs and outputs increase the numbers of dendrites and synapses within the rat brain. This kind of change is available to all species. However, special changes in humans like those just outlined, and not evident in other primates, could have much greater effects upon the human brain.

First, the level of external inputs and outputs in developing humans is enormously greater than that in other species. For example, a caressing, singing, and speaking mother has influences upon the brain and body of the baby in her womb that is much less available to apes. This is evidenced by changed patterns of movements of the fetus to touch and sound, and possibly, even, to maternal emotional state. Even a father may play a part here by caressing the mother's abdomen, and singing and talking to the child in the womb.

There is now evidence (from imaging studies) that the fetus indeed responds to such stimuli, and that these responses carry over in the newborn baby. Equivalent maternal and paternal influences continued on the infant and child, and likewise the slightly further distant influences of grandparents, siblings and others close to the family, even the effects of more distant community members, could have yet further effects on brain development in the infant and child. These cannot easily occur, or occur to much lesser degrees, in other species, even the great apes.

Such influences presumably affect all infant and childhood experiences and, if we extend the meaning of the term, can be called 'education'. This covers not just formal education as we generally recognize it, but the 'education' that stems from all influences, both before and after birth, even to a creature that cannot, at that point, speak, but can certainly feel, see, hear and communicate in its own way.

Similarly, new changes in brain organization may come through internal input/output effects on development, such as the molecular factors in the cascade of developmental processes that flow from genes during development. It is now well understood that such factors not only modify the instructions of the direct genetic blueprint, but also, through upstream and downstream effects, actually change the meaning of the genetic instructions themselves.

Such ideas were proposed theoretically long ago by Waddington (and explained through his epigenetic landscape, 1977). They have been re-introduced recently in books such as The Dependent Gene (Moore, 2001) and Nature Via Nurture (Ridley, 2003).

Let us put these concepts into pictures. Waddington's metaphor helps give clarity. Waddington described development from the zygote to the adult as the path that a ball takes rolling down a hill (Fig. 7.20). The original position of the ball at the hill-top represents the genome, the original genetic blueprint. As the ball rolls down the hill it describes the pathway of development and growth. This pathway depends upon both the starting position (whereabouts on the hillside, beginning blueprint, genetic information) and the contours of the hillside (the various downstream building processes, molecular and environmental factors, epigenetic factors as Waddington called them). The final position of the ball at the bottom of the hill represents adult form and function.

With this metaphor in mind, let us examine some new implications for the human brain. First, we will retreat from the full landscape as figured by Waddington and

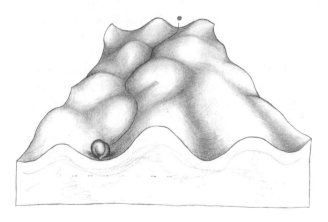

FIGURE 7.20

The epigenetic landscape (after Waddington).

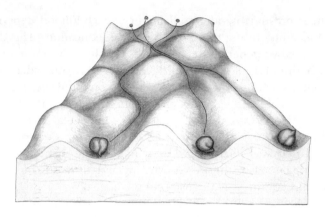

FIGURE 7.21

Small differences at the top of the hill (in the genome) together with intermediate differences in the developmental landscape can be expected to produce larger differences in the adult at the bottom.

move to simpler diagrams that nevertheless incorporate all the same features. Figure 7.21 shows how small differences at the top of the hill (in the genome) together with intermediate differences in the developmental landscape can be expected to produce larger differences in the adult at the bottom. At this point, our metaphor applies equally to other animals as to humans.

Let us next include the concept of population. Thus, Fig.7.22 shows the situation where there are populations of genomes. It shows dagrammatically how small population differences in genomes (slightly different starting positions of the ball) and somewhat bigger differences in the various epigenetic factors (somewhat bigger differences in the contours of the hillside at different points) could produce considerably larger differences in populations of adults (considerably larger differences in the final position of the balls). Again, in terms of materials for evolution, this does not imply any fundamental difference between humans and other forms.

Let us now include the effect of changes in the developmental landscape that are due to those contours that **are added by the organism (humans), because it (we) can** (Fig. 7.23). These might include some of those we described above: (i) the (now standard) effects of the interactions between the various internal upstream and downstream molecular factors and the genetic factors themselves; (ii) the effects of external factors such as sensory inputs, education, and so on (if we think we are clever enough to know what aspects of education actually improve learning); (iii) the complex and increasing effects of a much greater volume and activity of glia (if eventually proven), (iv) the improvements due to increased use of animal food-stuffs (e.g. better energy, animal protein, and increased micro-nutrients (e.g. vitamin B_{12} components)), but perhaps most of all (v) the interactions that might occur on the brain among all these different influencing modalities (i)–(iv).

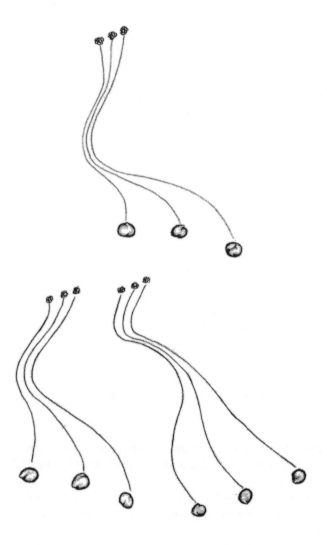

FIGURE 7.22

The epigenetic landscape where there are populations of genomes. Small population differences in genomes and somewhat bigger differences in epigenetic factors could result in considerable differences in populations of adults.

However, Fig. 7.23 also implies that the effect of all these factors might not only produce new forms of adults but also eliminate some adults that would otherwise have been produced. Could this last give rise to a division within the population, the creation of a class and an underclass? Of course, at the start, the genomes remain the same, and so one would not expect this to produce evolutionary change directly. It could result, however, in a well-known situation that might engender evolutionary change. That is, such a system might give rise to reproductive separations within

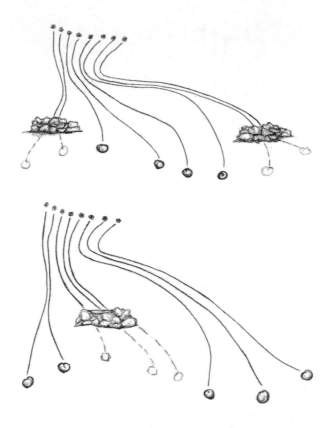

FIGURE 7.23

Effect of changes in the developmental landscape that may be produced by the organism – this may produce unification or separation of parts of the population.

populations. It is well known that this greatly speeds evolutionary separation. Of course, given the normal breeding propensities of humans this would seem, on the face of it, to be unlikely.

However, let us go even further. That is, this artificial interference in the contours of the developmental landscape might be carried out at earlier and earlier times during development (that is, higher up the hillside if you like (Fig. 7.24)). The resulting differences in the different adults might be greater and greater. The figure shows that it might result in a very much increased gap between adult groups, thus greatly increasing the chances of reproductive isolation. Earlier and earlier interpolation of education (in our widened view of the concept) into the developmental process might do this. Even so, the ordinary forms of education with which we are all familiar, are unlikely to produce sexual isolation.

But new types of inputs to the brain, engendered by the use of the computer and information technology, applied at increasingly younger ages, to increasingly limited portions of the population, just might do this. Who amongst us has not been embarrassed by the teenager, the child, even the infant, who knows how to 'work

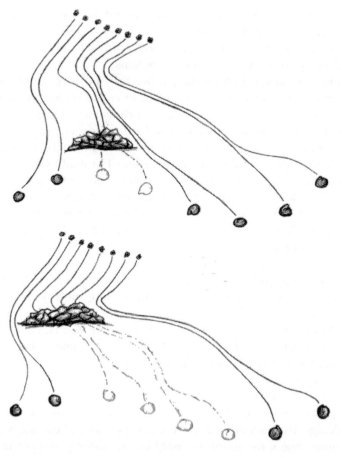

FIGURE 7.24

Effect of change in the developmental landscape (e.g. education) that may be produced by the organism. If carried out at earlier and earlier developmental times such change could produce greater and greater divisions within the population.

the machine' and who is scathing that 'nanny' and 'grandad' can't. It is just possible that this new information technology may do something very interesting to brains of the information generations of *Homo sapiens*. New non-invasive imaging methods might tell us.

Of course, this effect might also be strongly negative. We already hear voices warning about the dangers of computers to the brains of the very young. Uncontrolled access to computers and related gadgets by young children might well be strongly deleterious. This might result in *Homo nerdensis* and would seem to be an evolutionary liability.

Changes in other than computers might also be involved here. The separations that seem to be growing within many societies, the gaps between the haves and the have nots, the differences between the underfed and the rest, the greatly increasing monetary inequality in many societies, even in those societies with, ostensibly, the greatest resources, all these could lead to *Homo infra-classicus*, to societal disruptions, to societal splits, even to societal extinctions.

Changes might, however, be extremely powerfully positive; *Homo sapiens* might give rise to *Homo sapientior*; even, eventually, to *Homo sapientissimus*.

I doubt that *nerdensis* and *infra-classicus* individuals would very often breed with *sapientior* and *sapientissimus*; *nerdensis* and *infra-classicus* seem so often to suffer from such major behavioral problems in getting on with peers that breeding possibilities *in toto* might well be changed, reduced or might even not occur.

The reverse may also happen. *Homo sapientior*, even more so, *Homo sapientissimus* may well seek out only its own! Already improvements in women's education seem to be associated with women tending to choose only men who are educated at their own level or above. Are 'lesser' males being removed from the breeding pools of these 'super' women?

Are there really a number of new possibilities for the evolution of the human brain? But is this idea really new? Where have we seen it before? Was Julian Huxley (1942) correct in envisioning a new taxonomic group, the Psychozoa, containing, at this point, only *Homo sapiens*? Was Aldous Huxley (1932) correct in imagining his fictional Alpha-intellectuals and his Delta-minuses? Was H. G. Wells (both himself, 1895, and his alter ego, David Lake, 1981 almost a century later) percipient in envisioning *Eloi* and *Morlocks* in our human future?

Could *Homo* split?

Of course, this is unlikely.

It should, however, be absolutely clear that these ideas are nothing to do with long since denigrated eugenics or nasty social policies. These ideas do, however, make me wonder about the simplistic way human anatomy and evolution are often described by our discipline, and how they are presented in medias, encyclopedias, and wikipedias for the general public, and in the usual introductory texts for students (which this book is certainly not).

CHAPTER EIGHT

Postlude: Possible Human Futures

As we have seen throughout this book, the inferential study of human anatomy, the old method, has been greatly extended. Observation, measurement, and simple statistical analysis have become transformed by today's techniques: multivariate statistical analysis, geometric morphometrics, Fourier transforms, image analysis, and a host of others.

Equally, inferential guesses about human function have also been greatly extended by better understanding of what really happens and better examination of how it happens. Thus, to natural history descriptions of animal lifestyles have been added methods such as cinematography, cine-radiology and kinesiology to discover what really happens in nature. And the underlying mechanics uses techniques such as the

---

The Scientific Bases of Human Anatomy, First Edition. Charles Oxnard.
© 2015 John Wiley & Sons, Inc. Published 2015 by John Wiley & Sons, Inc.

older photoelasticity and strain gauge experiments, and the newer finite element analysis, Lagrangian functions, and more.

The results of such studies show us so often that what we expect to occur, does not, and that what we do not expect to occur, does. They therefore improve our knowledge of the functions of anatomies. They provide biological information relevant to the assessment of bones in evolutionary investigations. They even provide ideas relevant to clinical problems. A particularly fascinating feature in recent years has been how the arrangements of muscles, so clearly related to functional adaptation of bones as the motors of the body, have, as 'ghosts', gradually revealed new information about development and evolution (Oxnard, 2008). Of course, the new developmental biology helps to explain some of this anatomical information in the adult.

In a surprising reverse manner, however, the new anatomical information in the adult provides ideas, indeed predictions, for testing in the fetus, the embryo, the zygote, that is, for developmental biology. Anatomical units are most easily understood in the light of their functions, but the coalescence of anatomical units into whole organisms takes us into a whole new world of information. Function (though not lost) tends to slide out of sight. Evolution (present though previously hidden) stands revealed. The mechanism seems to be a reflection in adult structure of underlying environmental interactions, developmental processes, and hereditary units. This is the reverse of developmental biology which shows how the interaction of hereditary units, developmental processes, and environmental interactions produces whole organisms. It all depends upon the questions being asked.

General ideas on evolution then lead on to 'finding early ancestors', one critical part of the specifically human story. I do not mean by this the finding of fossils, but rather the assessing of the relationships of the fossils. In recent years the number of species recognized as being in the human part of the evolutionary tree has grown so enormously as to provide a human evolutionary bush. There are now (depending on how one names them) as many as 20 or more such species dating back over the past eight or even more millions of years. In contrast, there is only one specimen (and that named only a very few years ago) that has been placed in the otherwise totally 'branchless' part of the chimpanzee 'bush'. And there is nothing that is assessed as being neither a human ancestor nor an ape ancestor! These are unlikely scenarios!

Is it not really rather more likely that *Homo* as a genus has been around for very much longer than previously guessed? Is it not really rather likely that some of the wide range of fossils that are thought to lie in the evolutionary bush leading to modern humans actually lie in bushes leading to modern apes? Is it not most likely of all that some of the fossils actually lie on branches of the evolutionary bush that terminated in ancient extinctions?

Further, 'recognizing recent ancestors' seems to assume yet other improbable events. Could it really be the case that modern humans streamed out of Africa and replaced all earlier humans elsewhere in the world? Is it not really rather likely that modern humans, no doubt originally from Africa of course, have emigrated several times into the rest of the world, and have, moreover, become intermixed with other older forms of *Homo* also in the rest of the world? New ideas about Neanderthal genes being present in modern humans and about new versions of humans that are neither modern nor Neanderthal (e.g. Denisovans, and probably others will be found) support such a notion.

Finally, 'recognizing close relatives' seems to assume yet other improbable factors. So many current views of the relationships between chimpanzees and humans seem concentrated on the degrees of similarity between them. Is it not really more likely that the key is to look for degrees of differences? Changes in genes, in DNA, are probably the least important; they provide that 98.6% DNA similarity figure for humans with chimpanzees.

Changes in other aspects of development distal to the gene, in downhill developmental factors affecting how genes become expressed, in environmental features likewise affecting those downhill factors, in the interactions between genes and the environment, and above all, in changes in timings of factors, features, and interactions, are all, surely, far greater keys to differences between us and our closest living relatives. These differences are likely to be most evident in adults, and especially in adult anatomies. This may be especially true of the human side of the story where these differences can be greatly magnified in a creature that can, uniquely, affect changes in its own fetus, baby, child, youth, family, society, even possibly species (including species extinction – *vide* global warming, nuclear holocaust, meteoritic impact, black hole attraction, etc., whichever comes first).

When biological aspects of the body are examined, modern humans are indeed incredibly close to modern apes: same skin, same flesh, same blood, same bones. Most anatomies and most genes tell us this. There is not much difference between livers, kidneys and bone marrow in chimpanzees and humans; in fact, together, they differ much more from livers, kidneys and bone marrow in rhesus monkeys.

But, as we have seen, the human brain is special. There are enormous differences between brains in chimpanzees and humans, even between casts of internal skull surfaces (surrogates of the exterior of the brain) in chimpanzees and humans. Thus, in brains, in contrast to bodies, the similarity is between chimpanzees and rhesus monkeys. It is humans that are different: different brain structure organizations, different brain functions, different brain capabilities, different thoughts (as far as we can tell).

It is true that some of the complexities of human brains, especially their great size and the great sizes of their major components, represent just an extension of a trend common to non-human primates. This trend leads from monkeys to apes to humans. In such features, even though somewhat distant from apes, humans are just an extension from the ape condition, and they in their turn just an extension from the monkey condition.

As we have seen, however (Chapter 7), there are a series of other features of human brains, features involving complexities between major brain parts that differ from all non-human primate brains, even including ape brains, in new and unsuspected ways. Quantitatively, these differences of humans from chimpanzees are almost an order of magnitude greater than the differences between chimpanzees and rhesus monkeys. Here it is humans that are unique. Yet, this finding is based only upon study of major brain parts. There seems little doubt that if such studies could allow full consideration of interactions within major brain parts, especially within the cortex, the differences would be enormously greater still. Let us be immediately aware that this is not a return to an old, and ill-informed, idea: *pace* the hippocampus minor of a prior generation of scientists.

This brain uniqueness is supported by new data. Some of these data are developmental, involving at least four times as much difference in changes of human brain molecules, and possibly a great deal more, as compared with the relative stability of chimpanzee and rhesus monkey brain molecules. Some are comparative across the species with anatomical differences in various components of the brain in humans that are some 10 to 30 times greater from chimpanzees than from chimpanzees to rhesus monkeys. And some result from such comparisons over time as can be made from time sequences (early, middle and late) from skull data. Thus, the breadth of the frontal anterior cranial fossa, the depth of the middle cranial fossa, the vertical extent of the parietal, even and especially interestingly, the size of the cribriform plate of the ethmoid (these reflecting, to some degree, the sizes of frontal temporal and parietal cortices and the olfactory bulb) are all components related to the neocortex, and are all among the things that have changed most in the genus *Homo* and to by far the greatest extent in modern *Homo sapiens*.

These faster changes and increased complexities in the brain allow for new possibilities. Julian Huxley's view of taxonomy nostalgically wished to return to the old 'grades' of Protozoa and Metazoa, but with a new grade added, the Psychozoa, containing only humans. He wanted to recognize a whole new grade, with whole new modes of evolution essentially unique to humans. The taxonomic argument is an old one and not important. But perhaps recognition of new places and new mechanisms for humans is indeed newly important.

When read superficially, this is not a popular view. Many workers in human evolution imply that those features of humans that are especially associated with us being human, for example, our aggression, our social structure, our creativity, even our mysticism, (there are many other such categories) are closely related to prior ape characteristics such as ape aggression, ape social structures, ape creativity and precursors to ape mysticism (whatever they might be).

The differences of modern human behaviors resulting from these brain differences are, therefore, usually portrayed, to use a geological metaphor, as a very thin veneer, or slice, or stratum of human behavior superimposed upon a very thick multilayered bedrock of complex prior ape, prior primate, even prior mammalian, lifestyles, and that this is a measure of very close evolutionary relationships, and very short periods of time. This seems to be the viewpoint of many who look to apes and other primates, even other mammals, for the primary determinants of human behavior.

If, however, the new time relationships of the fossils (humans and chimpanzees now possibly three or four times more distantly related than used to be thought), if new velocity differences, for example, increased speeds of change in molecules of human brains (human brains now molecularly have changed at least three or four times more quickly than those of chimpanzees), and if elaborations in human brain organization (at least 10 to 30 times greater in structural complexity) are correct, then new possibilities open up.

If a brain with a beginning human-like volume (say 900–1100 cm^3 in *Homo erectus*) has existed for say 3 million years, rather than just a few hundred thousand years; if bipedality has existed for possibly even 3 or 4 million years; if bipedality of some form or other (not necessarily the same as that of modern humans) figured on the evolutionary scene as long as 5 or even 7 million years ago; if some kind of fossil and molecular transition from a common ancestor of apes and humans to an ancestor of humans alone really occurred at 8 or even 10 million years (all times permitted by different recent investigators and commentators); or even, if at least only some of these things occurred, then there will have been so much more time for the neural differentiation of human lineages from ape lineages.

Given also that it is likely that brain evolution in humans may have been proceeding at much greater rates than the more usual glacial bodily evolution, it may well be that enormous changes have taken place in prehuman and human lineages. It is especially likely, moreover, that the rate of this change, at first rather slow and therefore perhaps hard to discern, has been increasing by greater and greater degrees in shorter and shorter periods of time, becoming, therefore, more and more obvious. This last 2 million years, more, this last 200 000 years, even more, this last 30 000 years, even this last 10 000 years, even possibly in the short period of the growth of science, may have left not only our earlier progenitors far, far behind, but even also our more recent human-like parallels and/or progenitors. It is even possible that this change is still occurring now at speeds so great as to surprise us.

Given all this, some might think a bit of a tall order, but not at all impossible following the prior thinking in this chapter, rather than human abilities being a thin veneer upon a deep and complex bedrock of animal characteristics, we may totally rework the metaphor. We may think of the cognitive capacities of humans and the brain supporting them as an enormously thick complex of neural sludge overlying what is, in comparison, only a very thin underlying bedrock of prior animal behavior.

As we think of what it is that makes us human this idea does not seem so extreme. Although many investigators have tried to draw close analogies between human behaviors and ape behaviors, we must admit that, as common sense lay individuals, such efforts try our patience to the utmost.

Animal aggression, ape aggression, must presumably be the initial determinant of human aggression. Let us look, however, at the enormous complexities of human aggression today: nastiness of children to one another in the nursery and the playground; bullying in the family, the school and the workplace; road rage in car parks and on freeways; family violence: beating, not only of children, but of spouses and partners, of parents, and of grandparents; community violence: such as robbery, assault and battery, rape, rape with violence, rape even with murder, murder as a crime of passion, murder in the act of committing some other crime, murder in cold blood (as we say), sadistic and sexually oriented murder, violence and killing in the tribes, the religions, the nationalities and the internationalities; rape, gang rape, torture, mutilation, mayhem, murder and mass-murder all as tools of warfare, as a 'normal' way of carrying out some of the business of our 'civilisations', torture and murder to obtain information both in war and 'peace', killing through hand to hand combat, indiscriminate killing through terrorism, maiming and killing using child soldiers, killing using suicidal or brain-washed volunteers, killing through long-range weapons, through pilotless machines, push-button techniques, cyber killing, wholesale massacre, attempted and successful genocide, even a vision of the death of the world.

There is almost no limit to the aggressive ends of human behavior; it seems most unlikely that this is a mere thin veneer lying upon the thick bedrock of ape aggression.

Likewise, we may look upon the developments of our societies: our human laws, from the pronouncements of the shamans, the dooms of the kings, the tablets of the priesthood, the customs of the tribes, through the laws of the ancients, the complex forms and fictions of, especially, ancient Greece and Rome, through the court of Torquemada, the star chamber of British constitutional history, through trial by ordeal, punishment by burning at the stake, by hanging, drawing and quartering, the use of *peine forte et dure* as a method of 'persuading' a person to plead (later not even a punishment or persuader but accepted by a debtor for the relief of the family though it lead, of course, inevitably to the death of the debtor), to the extreme complex legalities of modern business, civil, criminal, international, sea and space law, even our new 'laws of ethics', and the use of the internet ('cybercrime') in interfering with the running of business, of cities, of countries, indeed of the world itself.

It seems highly unlikely that the thin layer of tentative personal interactions seen in ape social structures has much to do with these deep and complex developments in humans.

When, even further, we view the enormous complexities of human creativity, from the making and using of simple tools, through the design of implements that are not only useful but pleasing to the eye and the hand, through delight in our own home-made household articles, the potter's art, architecture, sculpture, painting, music, song and dance, through the creations of those artists and scientists who have been among the geniuses of humankind, the art of Michelangelo, the science of Einstein, the performance, more ephemeral but none the less real, of a Swan Lake ballet, the many other scientific, artistic, and humanistic creations of humans; when we look at all these things, it seems so very far from a monkey washing food in a pool or a chimpanzee shaping a twig to eat termites.

On such scores, precursor ape behaviors seem such a low base compared with the enormous heights of human creativity.

Even, finally, we may look at the ethical, moral, religious, and mystical creations of the human mind; even some of the perversions of these things, such as religious dogma, sectarian heresy, barbaric fundamentalism, torture, murder and terrorism in the name of 'churches', and 'religions', even sects, astrology, Satanism, black magic, the modern cults, and so on.

It seems that we cannot even begin to see in the thin precursor layers in apes, such thick and complex layers of activities, and thoughts, and visions as are present in humans.

An earlier generation of workers saw in these human characteristics developments that were nothing to do with apes and biological evolution, but were purely socially determined. A later generation of scientists saw in them a simple thin extension of the thick bedrock of genetically determined ape behaviors. Today many still emphasize the closeness of humans and apes, some even being willing to place them in the same genus! I agree that this last is a very important development when it is used to emphasize the preservation of species and of life. I support that as strongly as the rest of us.

But that last humanitarian ('animalitarian') viewpoint may just get in the way of understanding the science of what has really happened to humans in the recent past, what is still happening to humans today, and what in fact may be continuing to happen to us in the future.

Perhaps today we should really recognize a very complex situation, the result not just of 'genes' and 'environment', not just even of an interaction between genes and environment, but of the whole new cascades of many interacting molecular and environmental factors and timings. Perhaps we should look to an end, that is to say, of arguments about 'nature versus nurture', even about 'nature-via nurture' and recognise a much further explication of a complex interactive 'nature vis-à-vis nurture' permitted by new evolutionary mechanisms that do not occur in other animals, that are unique to humans.

For me, as a convinced (but modified) Popperian (Popper 1959), all this is predicated upon my genuine desire to use my thoughts and work to test current ideas about human evolution, to attempt to show them wrong, to thereby suggest better ideas for testing in their turn. My efforts are, however, a very far cry from the stupidities of 'intelligent design' and 'creation science'.

Pessimistically speaking, I cannot but expect that my continuing desire to test evolutionary ideas will be treated as support, as has been so much of my previous work, for these crazy fundamentalist beliefs.

Optimistically speaking, I hope to see human studies progress by adoption of an investigative stance that wishes to test all evolutionary ideas, even those that are most cherished, and as a result gain ever better understanding of the 'new' human condition.

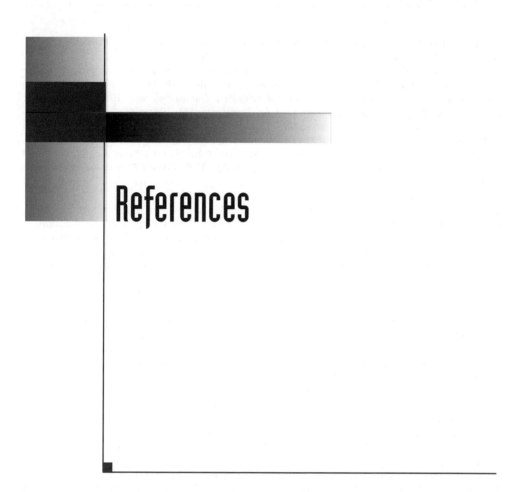

References

Arthur, W. (1997) *The Origin of Animal Body Plans*, Cambridge University Press, Cambridge.

Carlson, B.M. (2004) *Human Embryology and Developmental Biology*, 3rd edn, Elsevier Inc., Philadelphia.

Carter, D.R. (2001) *Skeletal Function and Form*, Cambridge University Press, Cambridge.

Carter, D.R. and Beaupré, G.S. (2001) *Skeletal Function and Form*, Cambridge University Press, Cambridge.

Cartmill, M., Hylander, W.L., and Shafland, J. (1987) *Human Structure*, Harvard University Press, Cambridge.

Cartmill, M. and Smith, F.H. (2009) *The Human Lineage*, Wiley-Blackwell, Hoboken.

Carey, N. (2012) *The Epigenetics Revolution*, Icon Books, London.

Crick, F. and Watson, J. (1953) Molecular structure of nucleic acids: A structure for deoxyribose nucleic acid, *Nature*, 171, 737–738.

The Scientific Bases of Human Anatomy, First Edition. Charles Oxnard.

© 2015 John Wiley & Sons, Inc. Published 2015 by John Wiley & Sons, Inc.

Currey, J. (1984) *The Mechanical Adaptations of Bones*, Princeton University Press, Princeton.

Darwin, C. (1859) *On the Origin of Species*, Murray, London.

de Winter, W. and Oxnard, C. (2001) Evolutionary radiations and convergences in the structural organization of mammalian brains, *Nature*, 409, 710–714.

Friedenthal, H. (1900) Ueber einen experimentellen Nachweiss von Blutzverwanschaft. *Archives of Anatomy and Physiology Leipzig, Physiology Abt.* 494–508.

Hennig, W. (1966) *Phylogenetic Systematics*, University of Illinois Press, Urbana.

Hall, B.K. (1999) *Evolutionary and Developmental Biology*, 2nd edn, Kluwer, Dordrecht.

Hildebrand, M. (1974) *Analysis of Vertebrate Structure*, John Wiley & Sons Inc., New York.

Hildebrand, M., Bramble, D.M., Liem, K.F., and Wake, D.B. (1985) *Functional Vertebrate Anatomy*, Harvard University Press, Cambridge.

Huxley, A. (1932) *Brave New World*, Chatto and Windus, London.

Huxley, J. (1942) *Evolution. The Modern Synthesis*, Allen and Unwin, London.

Huxley, T.H. (1863) *Evidence as to Man's Place in Nature*, Williams and Norgate, London.

Hyman, L. (1942) *Comparative Vertebrate Anatomy*, University of Chicago Press, Chicago.

Jerison, H.J. (1973) *Evolution of the Brain and Intelligence*, Academic Press, New York.

Kandel, E.R, Schwartz J.H, and Jessel T.M (2000) *Principles of Neural Science,* McGraw-Hill, New York.

Kardong, K.V. (2009) *Vertebrates: Comparative Anatomy, Function,* Evolution, McGraw-Hill, New York.

Keith, Sir A. (1902) *Human Embryology and Morphology*, Edward Arnold, London.

Lake, D.J. (1981) *The Man Who Loved Morlocks*, Hyland House, Melbourne.

Larson, W.J. (2001) *Human Embryology*, 3rd edn, Churchill Livingston, Philadelphia.

Lisowski, F.P. and Oxnard, C.E. (2007) *Anatomical Terms and their Derivation*. Student Pocket Book. *Vade Mecum,* L. (Latin), Go with me, World Scientific, Singapore.

Mayr, E. (2001) *What Evolution Is*, Basic Books, New York.

Moore, D.S. (2001) *The Dependent Gene. The Fallacy of "Nature vs. Nurture"*, Henry Holt, New York.

Nigg, B.M. and Herzog, W. (1994) *Biomechanics*, John Wiley & Sons Inc., New York.

Nuttall, G.H.F. (1904) *Blood Immunity and Blood Relationship*, Cambridge University Press, Cambridge.

Owen, R. (1868) *On the Anatomy of Vertebrates*, Longmans, Green, London.

Oxnard, C.E. (1975) The place of the australopithecines in human evolution: grounds for doubt? *Nature*, 258 (5534), 389–395.

Oxnard, C.E. (1983/1984) *The Order of Man: a Biomathematical Anatomy of the Primates*, 1983, Hong Kong University Press, Hong Kong, 1984, Yale University Press, New Haven.

Oxnard, C. (2008) *Ghostly Muscles, Wrinkled Brains,* Heresies and Hobbits, World Scientific, Singapore.

Oxnard, C.E., Crompton, R.H., and Lieberman, S.S. (1990) *Animal Lifestyles and Anatomies*, University of Washington Press, Seattle.

Page, R.D.M. and Holmes, E.C. (1998) *Molecular Evolution: A Phylogenetic Approach*, Blackwell, Oxford.

Popper, K.R. (1959) *The Logic of Scientific Discovery*, Routledge, London.

Ramachanandran, V.S. (2011) *The Tell-Tale Brain: A Neuroscientist's Quest for What Makes Us Human*, Norton, New York.

Ridley, M. (2003) *Nature Via Nurture: Genes, Experience and What Makes Us Human*, Harper Collins, London.

Shubin, N. (2008) *Your Inner Fish*, Vintage Books, New York.

Stern, J.T. Jr. (1988) *Essentials of Gross Anatomy*, Davis Co., Philadelphia.

Stern, J.T. Jr. and Oxnard, C.E. (1973) *Primate Locomotion: Some Links with Evolution and Morphology*, Karger, Basel.

Striedter, G.F. (2005) *Principles of Brain Evolution*, Sinauer, Sunderland.

Thompson, D'A.W. (1917) *On Growth and Form*, Cambridge: University Press, Cambridge.

Todd, R.B. and Bowman W. (1845) *The Physiological Anatomy and Physiology of Man*, Parker, London.

Twyman, R.M. (2001) *Developmental Biology*, Springer, New York.

Waddington, C.H. (1977) *Tools for Thought*, Basic Books, New York.

Wainright, S.A., Biggs W.D., Currey J. D., and Gosline, J. M. (1976) *Mechanical Design in Organisms*, John Wiley & Sons, New York.

Wells, H.G. (1895) *The Time Machine*, Heinemann, London.

Wells, H.G. (1933) *The Shape of Things to Come*, Hutchinson, London.

Wiedersheim, R. and Parker W.N. (1907) *Comparative Anatomy of Vertebrates*, 3rd edn, Macmillan, London (based on 1882 and later editions of Wiedersheim, R.. *Lehrbuch der Vergleichende Anatomie der Wirbeltier*).

Wiedersheim, R. (1895) *The Structure of Man: An index to his Past History*, (Transl. H. and M. Bernard, ed. G. B. Howes), Macmillan, London.

Young, J.Z. (1950) *The Life of Vertebrates*, Clarendon, Oxford.

Young, J.Z. (1966) *The Life of Mammals*, Clarendon, Oxford.

Young, J.Z. (1971) *An Introduction to the Study of Man*, Clarendon, Oxford.

Zuckerkandl, E. and Pauling, L. (1962) Molecular disease, evolution, and genic heterogeneity, in *Horizons in Biochemistry*. (eds M. Kasha and B. Pullman, Academic Press, New York, pp. 189–225.

Zuckerman, S. (1933) *The Functional Affinities of Man, Monkeys and Apes*, Kegan Paul, London.

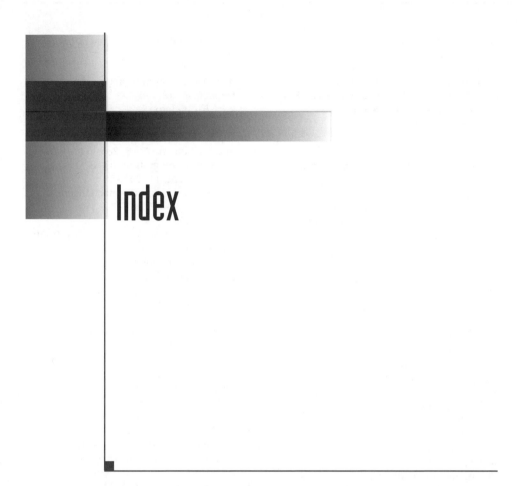

Index

The Scientific Bases of Human Anatomy, First Edition. Charles Oxnard.
© 2015 John Wiley & Sons, Inc. Published 2015 by John Wiley & Sons, Inc.